THE CASE AGAINST DISCLOSURE

DEFENDING CREATIVE AUTONOMY IN THE AGE OF AI

THE CASE AGAINST DISCLOSURE

DEFENDING CREATIVE AUTONOMY IN THE AGE OF AI

JAMES HUTSON

DANIEL PLATE

COMMON GROUND

First published in 2025
as part of the **Information, Medium & Society Book Imprint**

Common Ground Research Networks
University of Illinois Research Park
2001 South First St, Suite 201 L
Champaign, IL 61820 USA

Library of Congress Cataloging-in-Publication Data

Names: Hutson, James author | Plate, Daniel author
Title: The case against disclosure : defending creative autonomy in the age of AI / James Hutson, Daniel Plate.
Description: Champaign, IL : Common Ground Research Networks, University of Illinois Research Park, 2025. | Includes bibliographical references and index.
Identifiers: LCCN 2025013777 (print) | LCCN 2025013778 (ebook) | ISBN 9781966214564 hardback | ISBN 9781966214571 paperback | ISBN 9781966214588 pdf
Subjects: LCSH: Authorship—Technological innovations | Artificial intelligence—Philosophy | Creation (Literary, artistic, etc.)—Philosophy | Technology and the arts
Classification: LCC PN171.T43 H88 2025 (print) | LCC PN171.T43 (ebook) | DDC 808.0285—dc23/eng/20250530
LC record available at https://lccn.loc.gov/2025013777
LC ebook record available at https://lccn.loc.gov/2025013778

ISBN: 978-1-966214-56-4 (HBK)
ISBN: 978-1-966214-57-1 (PBK)
ISBN: 978-1-966214-58-8 (PDF)
DOI: 10.18848/978-1-966214-58-8/CGP

DEDICATION

This manuscript is dedicated to our families, whose unwavering support and encouragement have been the cornerstone of our journey. To our parents, we express deep gratitude for instilling in us a thirst for knowledge and the determination to pursue our passions. Your belief in our abilities and unwavering support have propelled us forward.

To our spouses and partners, we extend heartfelt appreciation for standing by our side throughout the long hours of research and writing. To our children, we are grateful for your inspiration and the joy you bring to our lives. Your presence reminds us of the importance of our work and motivates us to strive for excellence. To our extended families, friends, and loved ones, we thank you for your encouragement, words of wisdom, and unwavering belief in our abilities. In particular, James Hutson would like to thank his wife Piper and children Bishop and Aurora.

Your support has been a source of strength and inspiration. We acknowledge the sacrifices you have made and the understanding you have shown, allowing us the time and space to pursue our scholarly endeavors. This manuscript stands as a testament to your love and support. With deepest appreciation and heartfelt gratitude, we dedicate this work to our families.

EPIGRAPH

"The creative process is a process of surrender, not control."
—Julia Cameron

ACKNOWLEDGMENTS

The authors express profound appreciation to the myriad individuals and institutions whose contributions and unwavering support have been pivotal in the genesis and fruition of this tome. At the vanguard of acknowledgment stands Dr. John Porter, whose innovative leadership in aligning the adoption of emergent technologies like AI with educational initiatives has left an indelible mark on this work's evolution.

Acknowledgment is due to Dr. Kathi Vosevich, Dean of the College of Arts and Humanities, for her steadfast encouragement and support of the research culminating in this publication. Vosevich's guidance and confidence in the authors' academic endeavors have been fundamental in sculpting the contours of this volume.

The authors extend their gratitude to the research team responsible for composing this manuscript and acknowledge the critical insights and contributions of various colleagues and coauthors from ancillary projects. James Hutson extends thanks to Professors Ben Fulcher, Jeremiah Ratican, Joe Weber, Ben Scholle, and Dr. Trent Olsen. As well, Drs. Daniel Plate and Emily Barnes were instrumental in shaping many of the sections and the direction.

The authors also convey their deep appreciation to the administrative leaders at all of the institutions that participated in the study for their relentless support and dedication to creating an environment conducive to academic excellence and scholarly exploration. The leadership, encompassing Presidents, Provosts, and Deans at these institutions, has been instrumental in providing the necessary guidance and resources for this research and writing project.

To those who provided insight into their experiences as leaders in industry and education, we are indebted and thank you for your invaluable time and efforts in helping others navigate this unpredictable time. Finally, the authors are thankful for the insights and constructive feedback from colleagues and fellow researchers throughout this project. Their contributions have significantly enriched the work and helped to refine the ideas encapsulated within this volume.

TABLE OF CONTENTS

ABOUT THE AUTHORS

Dr. James Hutson specializes in multidisciplinary research that encompasses artificial intelligence, neurohumanities, neurodiversity, immersive realities, and the gamification of education. Earning a Bachelor of Arts in Art from the University of Tulsa, a Master of Arts in Art History from Southern Methodist University, and a PhD in Art History from the University of Maryland, College Park, he later acquired additional Master's degrees in Leadership and Game Design from Lindenwood University and additional PhD in Artificial Intelligence at Capitol Technology University (2023). Over the span of his academic career since 2006, Hutson has held various pedagogical and administrative positions across five universities, including Chair of Art History, Assistant Dean of Graduate and Online Programs, and most recently, Lead XR Disruptor and Department Head of Art History, AI, and Visual Culture. Notably, his scholarly portfolio includes several books on the application of artificial intelligence in education and cultural heritage, such as *Creative Convergence: The AI Renaissance in Art and Design* (2024), as well as numerous articles and case studies.

Dr. Daniel Plate, a preeminent scholar in the field of educational technology, focuses his research and pedagogy on the integration of generative artificial intelligence in pedagogical methodologies. Educated with a bachelor of arts in English and philosophy from Taylor University, a master of fine arts in creative writing from the University of Arkansas, and a PhD in literature from Washington University in St. Louis, Plate presently serves as a faculty member at Lindenwood University. There, he instructs courses in creative writing, literature, and composition. Beyond his contributions to poetry, he has coauthored numerous case studies that explore the nuanced applications of AI in classroom settings. These scholarly works not only contribute to the existing body of literature but also offer practical insights for educators interested in leveraging generative AI to enrich the learning experience. Plate's scholarly endeavors straddle the intersection of technology and pedagogy; he divides his professional time between developing code to augment teaching methods and conducting empirical research to better understand the symbiotic relationship between AI and innovative pedagogy. A prolific author in the field of AI in education, his latest publication is *Generative AI in the English Composition Classroom: Practical and Adaptable Strategies* (2024).

LIST OF FIGURES

LIST OF TABLES

PREFACE

Each epoch perceives its new technologies as uniquely terrifying. When writing first emerged as a revolutionary tool, Socrates famously condemned it as the destroyer of memory itself. Ironically, scholars remember Socrates' disdain precisely because Plato immortalized his words through writing. Centuries later, Gutenberg's printing press provoked an equally intense upheaval, unleashing a flood of affordable knowledge that sparked revolutions, eroded entrenched power structures, and, most distressingly for some, rendered monks' painstaking manuscript illumination virtually obsolete. One can only imagine the horror on their ink-stained faces when confronted by movable type—an invention poised to replace handcrafted beauty with mechanical reproduction. Yet, civilization marched (or printed) onward, unfazed by monkish despair.

Fast-forward a few more centuries—the advent of personal computing revolutionized the physical act of writing, while the Internet irrevocably altered how information is created, shared, attributed, and monetized. Recall, if you will, the initial outrage directed at Wikipedia—"Anyone can edit it!" Critics warned of intellectual anarchy, yet today, Wikipedia seems utterly ordinary, even indispensable. Now, the latest focus of scholarly anxiety is generative artificial intelligence. Technologies such as ChatGPT's eerily persuasive prose and DALL-E's unsettlingly realistic images provoke existential unease across academia, traditionally the guardian of intellectual rigor and methodical inquiry. How can scholarly institutions trust work produced by invisible computational assistants? What ethical imperatives, if any, obligate creators to disclose precisely how their ideas and texts came to fruition? And perhaps most provocatively, does such transparency truly matter?

This book argues that demands for full transparency regarding AI-driven methods are not only practically untenable but fundamentally incompatible with creativity itself. The expectation that creators should meticulously document their every inspiration and technological assistance defies historical norms. Shakespeare never felt compelled to disclose whether Marlowe whispered narrative suggestions over pints of ale; nor did Monet append footnotes clarifying which brushstrokes emerged from conversations with Renoir. Credibility in creative works historically rests not upon exhaustive documentation of methods but upon the creator's willingness to claim responsibility and ownership of the final product—no matter how obscure or collaborative its origins.

Nonetheless, dismissing the profound disruption caused by generative technologies would be disingenuous. These tools are reshaping industries and academic disciplines at unprecedented speed. Within mere months, generative AI has shifted from curious novelty to essential productivity aid in many professional settings. Academia, a bastion of traditionalism and cautious adaptation, vigorously resists this shift, urgently erecting standards designed to delineate human from machine, authentic from artificial. Such standards, however noble in intent, collapse under practical scrutiny. Exactly how detailed must disclosures become to satisfy demands for transparency? Should creators retroactively disclose the influence of their word processors, their grammar-checking software, or even—heaven forbid—Google?

Indeed, there are two dominant reactions underpinning contemporary resistance to generative AI. The first is economic—a legitimate fear that these intelligent tools threaten livelihoods by replacing jobs traditionally held by humans. Historically, each technological shift—from mechanical looms to automated factories—has provoked similar economic anxieties, with societies fretting over unemployment and the displacement of skilled labor. Today, writers, scholars, and artists understandably fear that their roles could be rendered obsolete by AI capable of producing not only serviceable but astonishingly competent creative content. Recently, reactions to a short story generated by AI illustrate this vividly. The exaggerated outrage online over its quality obscured the truly remarkable achievement that the story itself, while perhaps not Pulitzer-worthy, easily matches or exceeds the caliber of much fiction routinely published in respected literary journals. For someone who has frequently reviewed student creative writing, this development is truly astonishing: AI-generated fiction already rivals—and sometimes surpasses—human-produced narratives, a feat both unsettling and extraordinary.

The second reaction is rooted deeply in aesthetics and identity, tied intimately to a Romantic conception of art that most of us still hold. Since roughly 1770, Western society has largely lived within Romanticism's shadow, treating art as a quasi-religious pursuit, with the artist as a secular priest mediating profound spiritual truths. Thus, when confronted with competent AI-generated narratives, many readers instinctively recoil, insisting, "As soon as I know this was written by AI, it means nothing to me. I read literature because I want to read what humans express." This statement reflects a deeply ingrained belief that true art must express the unique, irreducible essence of human experience. Modern readers rely on this Romantic ideal, perceiving art as a precious link to a transcendent identity—especially crucial in a secular age where traditional spirituality has waned.

Yet the emergence of AI-generated art profoundly challenges this worldview. Can a machine truly produce art? Does it possess a soul or spiritual depth? The answer for most contemporary audiences is an emphatic "no." Such resistance reveals an anxiety not just about aesthetics but about losing a sense of human uniqueness itself. Interestingly, this fear might be uniquely modern. Would medieval craftsmen building grand cathedrals have objected to a machine capable of assembling precisely cut stone? Likely not, since their worldview did not privilege individual human creativity above collective, practical outcomes. By contrast, modern readers imbue art and literature with nearly religious significance, viewing it as humanity's highest form of self-expression. Hence, a competent AI-generated story isn't merely unsettling; it's perceived as sacrilegious.

Consider further questions raised by this aesthetic anxiety: Are genealogies in the Bible art? Are centuries-old sermons art? Historically, these texts served functional, community-oriented purposes, without necessarily emphasizing individual creative expression. Medieval Christians likely wouldn't object if an AI crafted a competent sermon or psalm, provided it served communal worship effectively. Yet contemporary believers, heavily influenced by Romantic notions of personal expression, would likely recoil at the suggestion. Our modern, secularized spirituality, which places art at its center, makes the new machine competence profoundly threatening.

In truth, current anxieties surrounding generated content merely echo historical responses to previous technological transformations—initial panic, reluctant adaptation, eventual normalization, and inevitable amnesia. The seemingly radical nature of generative technologies today will soon become mundane, as commonplace as emails, cloud computing, or indeed, Wikipedia itself. Academia's insistence on transparency, with hindsight, may appear as quaintly reactionary as monks demanding printed texts be meticulously hand-copied to preserve authenticity.

Consider the introduction of photography in the nineteenth century, which prompted painters to proclaim the imminent death of their art form. Mechanical reproduction, they feared, would eradicate originality and diminish artistic value. Yet painting survived—thrived even—evolving into Impressionism, Cubism, and Abstraction. New technologies have repeatedly confronted creative traditions, initially provoking alarm before quietly becoming integral to the creative toolkit. Today, generative AI confronts society with similar dilemmas, triggering contentious debates on authorship, authenticity, and creative boundaries. Demands for disclosure—clarifying precisely where human creativity ends and machine-generated contributions begin—reflect unprecedented standards fundamentally at odds with the messy, mysterious nature of creation itself. Creativity is rarely linear, neatly documented, or easily explicable; demanding such transparency misunderstands its very essence.

Admittedly, traditional artists, writers, and scholars will likely disagree with this book's central premise, and understandably so. Valid reasons persist for choosing traditional, technology-free methods of creation, just as legitimate motivations remain for immersing oneself in novels rather than transactional coding manuals. This book does not dismiss these choices; it merely recognizes that motivations and methods for reading and writing are being irrevocably re-shaped. Rather than outright rejection, the authors invite thoughtful dialogue and measured discourse. Agreement may remain elusive, yet consensus or respectful disagreement is urgently necessary. It is time to acknowledge hard truths, adapt pragmatically, and chart a considered path forward.

INTRODUCTION

Beyond Transparency

When Johannes Gutenberg's movable type clattered onto the European stage in the mid-fifteenth century, the uproar among traditionalists wasn't about the beauty of print but the peril of counterfeit texts — after all, how could anyone trust a book not painstakingly hand-copied by monks? Fast forward to our algorithmic age, and the narrative feels strikingly familiar: Anxieties about authenticity and originality once directed at print now orbit artificial intelligence (AI) tools like ChatGPT, with critics calling for meticulous transparency and disclosures as if creativity itself could be reduced to a recipe card. This chapter, therefore, examines the paradoxical demand that artists and authors lay bare their every prompt, click, and algorithmic adjustment, a call both historically unprecedented and fundamentally at odds with the elusive nature of creative impulse. Through a journey from Gutenberg's type blocks to the invisible hands behind large language models (LLMs), the analysis underscores how creativity, by definition resistant to complete transparency, inevitably resists attempts to audit its origins. In short, the real test of authorship isn't found in a detailed log of methodological minutiae but in the creator's courage to claim responsibility — whether the muse was ink-stained fingers, whispered prompts, or, heaven forbid, a chatbot's suggestion.

I.1 The Creative Process and the Myth of Disclosure

A war is raging, not on battlefields marked by blood and territory, but across fields of intellectual and creative endeavor — where artists, writers, poets, designers, researchers, and coders now grapple with an unprecedented perceived threat:

generative artificial intelligence (generative AI). On the surface, this conflict revolves around familiar anxieties, framed in reassuringly binary terms of human originality versus machine imitation. Skeptics argue vehemently that algorithms lack the ineffable human spark necessary to achieve true creativity, dismissing AI-generated works as inherently derivative or mechanistic. Beneath this ostensibly straightforward critique lies a deeper dispute—one about transparency and control. Indeed, calls to mandate disclosure of AI involvement signal an institutional urge to police creativity, dictating how creators should document their processes and reveal every mechanized influence. Yet such demands misunderstand the messy, idiosyncratic nature of creative work: They fail to recognize that artistic power derives not from total visibility, but precisely from its hidden, even inscrutable origins. This book contends that these mandates, though appearing ethical or protective of artistic authenticity, serve primarily as tools of hegemonic control that encroach upon the autonomy historically granted to creators. Far from preserving authenticity, disclosure mandates ironically threaten the creative processes their proponents claim to protect.

The terrain of AI-generated creativity has become increasingly fraught since the public launch of major generative platforms in the later half of 2022, marked by shifting guidelines, inconsistent regulatory stances, and widespread confusion among creators and institutions alike. Initial reactions oscillated wildly between unbridled enthusiasm and stark apprehension, as artists and writers rapidly embraced new tools such as Midjourney, Stable Diffusion, and ChatGPT, leading institutions and regulatory bodies to scramble for an appropriate response. One of the most contentious battlegrounds quickly emerged in copyright law, where the United States Copyright Office initially drew hard lines against AI-generated works, referencing *17 U.S.C. § 102*, deeming them insufficiently "creative" to merit legal protection. Kris Kashtanova's 2022 AI-assisted graphic novel, *Zarya of the Dawn* (Figure I.1), vividly illustrates this tension: Copyright protection was initially denied to the images generated by Midjourney but granted to the human-crafted narrative and structural design. This decision underscored a critical distinction—human storytelling and arrangement were protectable, machine outputs were not—affirming human oversight as central to legal authorship (Klukosky & Kohel, 2024). Yet, just as creators began adjusting to this precedent, another landmark ruling arrived, overturning earlier assumptions. In 2025, Kent Keirsey, CEO of Invoke, successfully obtained copyright protection for his AI-generated artwork, *A Slice of American Cheese*. However, securing this legal victory required demonstrating an exhaustive level of human involvement: Keirsey meticulously documented thirty-four distinct interventions, each

Figure I.1: Kris Kashtanova, *Zarya of the Dawn Cover*, Comic Book, 2022 (CC 0)

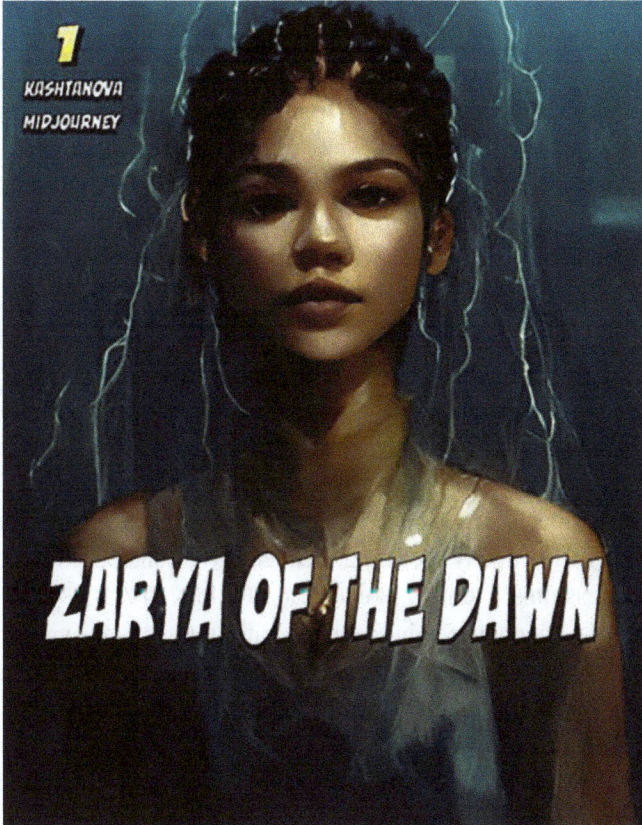

representing specific instructions issued directly to the AI (Growcoot, 2025). Ironically, each intervention was merely verbal—no direct brushstrokes or pen strokes were recorded—further complicating the already nebulous definition of "human creativity."

I.1.1 Creativity and the Art of Losing Control

Given this historical complexity, one might reasonably ask: What, precisely, is meant by the "creative process"? Defining creativity is notoriously elusive, a point underscored repeatedly throughout intellectual history. Plato's (428/427–348/347 BCE) dialogue *Ion* provides an instructive—if caustically amusing—illustration of this ambiguity. When interrogating Ion, a professional reciter of Homeric poetry, Socrates ridicules poets for their inability to articulate the origins of their ideas, suggesting inspiration arises mysteriously rather than through conscious, deliberate control (Dorter, 1973). Renaissance thinkers similarly grappled with this concept, notably articulated by Marsilio Ficino (1433–1499), who characterized poetic inspiration as *poetic furor*—a divinely inspired frenzy beyond rational explanation or conscious control. The "divine" Michelangelo (1475–1564) himself, for instance, described his own creative experience as a form of possession, declaring that his sculptures already existed within the marble; his task was merely to free them (Britnell, 1989).

In modern times, Surrealism further dramatized this belief through intentionally provocative practices: Salvador Dalí (1904–1989), for example, famously harnessed his subconscious by positioning himself to awaken abruptly from near-sleep states, jolted into awareness by a loud noise as an object slipped from his hand (Jiménez et al., 2013). Other Surrealists embraced automatic writing, creating texts spontaneously, without deliberate conscious intervention; others embraced experimental techniques such as the exquisite corpse method, assembling creations without any predetermined narrative coherence (Foster, 2014). Abstract expressionist painters like Jackson Pollock (1912–1956) employed action painting—random-seeming yet deliberate gestures—to bypass conscious intentionality altogether (Taylor et al., 2002). Even celebrated writers have historically sought to circumvent rational control: From Charles Baudelaire's (1821–1867) chemically induced poetic reveries to Ernest Hemingway's (1899–1961) liquor-infused prose, creators have frequently embraced the deliberate diminishment of conscious authority as integral to artistic authenticity. Ironically, contemporary critics who insist on strict transparency in AI-generated content overlook how historically central such opacity and unconscious spontaneity have been to creative authenticity. Yet now, in a baffling reversal, society demands creators precisely document—and thus control—their use of generative tools, imposing clarity where, traditionally, mystery and ambiguity have prevailed.

In fact, what precisely constitutes creativity has perplexed philosophers, artists, and scholars across history, raising fundamental questions about whether creativity truly originates within individual genius or instead emerges from ambiguous and inscrutable processes resistant to easy categorization or documentation. According to Merriam-Webster, creativity is "the ability to create," with the verb "create" embodying diverse actions and behaviors leading to something genuinely novel. Creating encompasses the deliberate and imaginative act of bringing ideas, forms, or expressions into existence, exemplified by everything from the biblical account of divine creation to the intricate processes of a fashion designer crafting an original garment. Darshan (2014) observes that, broadly considered, creativity bestows existence itself, infusing fresh roles and purposes through imaginative prowess. Such actions highlight human ingenuity, emphasizing the power to conceive, innovate, and materialize novel ideas and outcomes (Candy, 2002). As the figurative heart of artistic expression, creativity prompts individuals to explore beyond established knowledge and conventions, pushing the boundaries of innovation. Traditionally, as Lang (2019) emphasizes, the agency of the artist is pivotal; the creator consciously transforms raw, unstructured material into tangible expressions through deliberate engagement with their medium and the environment. Historically, this transformative engagement involved extensive mastery of tools and media, from Paul Cézanne's methodical explorations of form and composition (Smith, 2013) to the comprehensive artistic training detailed by Whitford (1923). Such mastery required considerable time and discipline, blending theoretical understanding with the rigorous development of practical skills (Stankiewicz, 2000). Musicians produce sound through their instruments; potters shape clay; authors craft narratives word-by-word. This educational and creative model was built on physical manipulation, the artisanal craftsmanship that defined creativity throughout history.

Yet, generative AI, marked by the explosive mainstream adoption of platforms like ChatGPT, Claude, Stable Diffusion, Midjourney, and DALLE-3 in late 2022, has disrupted this traditional paradigm profoundly. These tools challenge entrenched notions of authorship and control, shifting creativity from solitary genius toward collaborative innovation, as foretold by Somaya and Varshney (2020). Now, human artists and algorithmic partners work in tandem, utilizing machine capabilities to explore novel creative solutions (as in *Zarya of the Dawn* and *A Slice of American Cheese* above), while creativity permeates diverse fields beyond art and literature, including science, engineering, and business (Martiniano, 2016). Society, however, particularly venerates visual artists and authors due to their ability to translate abstract ideas into aesthetic or literary forms. The advent of AI alters this dynamic significantly. As Elton (1995) presciently noted, computational systems automate tasks traditionally demanding extensive human skill and training, reshaping educational practices for creatives. Ng (2021) similarly argues that mastering creative tools now means learning to communicate effectively with AI systems, leveraging algorithmic potential rather than manually manipulating physical materials. This shift, far from simplifying the creative process, expands it, providing unprecedented opportunities to explore forms and techniques previously impractical or impossible to realize.

This changing relationship between AI and creativity finds context in computational creativity's historical roots. Pioneers like Vera Molnar (Figure I.2) harnessed algorithmic methods decades prior to today's generative AI explosion. In literature, early AI experiments include works such as *The Policeman's Beard Is Half Constructed* (1984), crafted by the program Racter. Du Sautoy (2019), in "The creativity code," argues that AI's real promise lies in offering innovative inspiration and facilitating unprecedented artistic exploration. Both Ng (2021) and Elton (1995) contend that AI possesses creative agency beyond mere automation, generating outputs that surprise and challenge human collaborators. Such dynamics illustrate Boden's (1978, 1994) tripartite model of creativity: exploratory, transformational, and imaginative. According to this theory, AI excels in the first two domains—processing data, identifying patterns, and innovatively recombining elements—enhancing human creative practices. However, AI cannot replicate the uniquely human imaginative creativity that arises from personal intention, emotion, and meaning-making (James & Dewey, 2005; Manu, 2024). The degree to which this has held true (or not) will occupy much of our analysis.

Figure I.2: Vera Molnar, *Interruptions à recouvrements*, 1969 (CC 1.0)

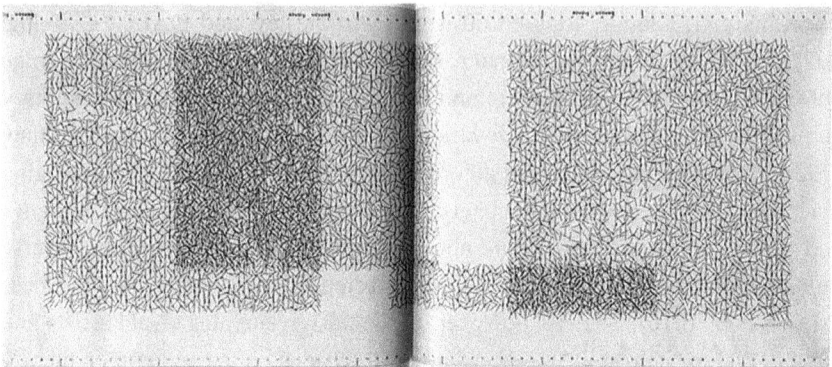

I.1.2 Mind over Matter: Prioritizing Concept over Craft

The tension between technological disruption and artistic tradition is not new. In the nineteenth century, the daguerreotype (*Daguerre Atelier*, 1837, Figure I.3) and other photomechanical reproduction technologies challenged traditional

Figure I.3: Mandé Daguerre, *Daguerreotype Daguerre Atelier*, 1837 (CC 0)

painting's representational monopoly. Painters faced existential questions about their craft, prompting the emergence of *avant-garde* movements that deliberately diverged from realism and objective depiction. Mitter (2008) emphasizes that technological disruptions consistently provoke redefinitions of creativity and the cultural roles of art. Similarly, in literature, mechanical reproduction via printmaking technologies challenged hand-copied manuscripts, prompting literary forms and storytelling methods to adapt profoundly. The anxieties provoked by contemporary generative technologies echo these historical moments, emphasizing that cultural anxiety about creativity's authenticity is neither new nor insurmountable.

For example, the most significant ideological break regarding what constitutes art emerged over a century ago with Marcel Duchamp (1887–1968) and the provocations of the Dada movement fundamentally challenging established definitions and expectations. Duchamp famously disrupted conventional norms in 1917 with the submission of *Fountain* (Figure I.4) — a commonplace porcelain urinal signed under the pseudonym "R. Mutt" — to an esteemed art exhibition. Rather than being an intricately crafted sculpture or painting,

Figure I.4: Marcel Duchamp, *Fountain*, 1917. Photograph by
Alfred Stieglitz (CC 0)

Duchamp's gesture emphasized conceptual intent over manual craftsmanship, prioritizing the intellectual act of designation above physical creation itself (Rudinow, 1981). Yet, as Goldsmith (1983) astutely observes, even Duchamp's radical ready-mades, including bicycle wheels and snow shovels, remained tethered to human experience, evoking familiar cultural meanings. Thus, traditional understandings of art — firmly rooted in human emotion, reflection, and expression — persist, reinforcing the association between creativity and humanity.

Given contemporary technological disruptions, however, it becomes increasingly pertinent to reconsider whether the boundaries of art and literature extend exclusively within human contexts. If creativity inherently implies bringing something new into existence, assigning it fresh forms or functions, and manifesting imaginative skill (Darshan, 2014), could the removal of the direct human touch still yield something worthy of designation as art? Immanuel Kant's (1724–1804) exploration of aesthetics suggested precisely such an expansion: The appreciation of beauty, according to Kant, occupies a mental space detached from rational explanation, thus allowing humans to appreciate natural phenomena—sunsets, roses, oceans—as inherently beautiful without human intervention. Danto (2013) notes that Kantian aesthetics, emphasizing this "disinterested" contemplation, has profoundly shaped modern definitions of art as detached from explicit religious, economic, or political functions. Hamilton (2024) furthers this argument, highlighting that such an expanded, inclusive approach embraces various creative expressions and broadens traditional limits beyond mere practicality or commercial utility. Indeed, this inclusive perspective aligns closely with contemporary artistic trends. Grundberg (2021) observes how the art world today increasingly explores unconventional materials and cross-disciplinary practices, transcending traditional venues like galleries or canvases through installations, digital media, performances, and conceptual forms. The legacy of performance art, Happenings, and Conceptual Art has significantly blurred distinctions between creator and observer, reinforcing the Barthesian notion of the "death of the author," which shifts interpretative authority toward the viewer. This recognition further expands the realm of creativity beyond traditional human frameworks, opening avenues for reevaluating and appreciating varied incarnations of art.

With generative AI now firmly established, the role of human input within creative outputs undergoes another profound recalibration. Generative models such as Stable Diffusion by Stability AI have been trained extensively on human-created content, using vast repositories like the LAION-5B dataset—a collection of over 5 billion human-generated image–text pairs from the web. Consequently, AI-produced images remain inextricably linked to human artistic and cultural traditions. The same principle holds true for textual generative models, whose linguistic outputs depend upon absorbing vast corpora of human-authored text, illustrating the phenomenon of "Absent Presence," wherein human influence permeates algorithmic results implicitly yet undeniably. Certainly, generative AI introduces innovative and insightful rearrangements of existing knowledge, capable of yielding novel perspectives and new artistic possibilities. Nonetheless, as Lang (2019) emphasizes, converting creative inspiration into tangible forms remains central to art and design. Whether an artist transforms an AI-generated visualization into a physical artwork, or a novelist crafts narratives inspired by algorithmic prompts, the creative process fundamentally parallels traditional studio practices, wherein photographs or sketches serve as points of departure. Thus, despite superficial appearances of reduced human involvement, machine creativity inherently retains traces of human ingenuity, intention, and cultural context. This collaborative framework positions generative tools as a potent creative partner, capable of extending human imagination and enriching creative practice. Yet, as Manyika (2022) points out, generative algorithms remain limited in emulating the emotional depth, nuanced understanding, and empathy central to meaningful human expression. These profoundly human qualities continue to underpin the most compelling artistic and literary creations.

Still, while there is a degree of continuity in human creativity, generative AI should not be seen as merely another incremental invention; it marks a profound shift in how creativity and authorship are conceptualized, shaking traditional frameworks to their foundations. Indeed, proclaiming the demise of human creativity due to technological innovation ignores historical lessons about adaptation and the persistent resilience of creative impulses. Yet this latest disruption does represent a significant departure. The impending transformation is not simply another incremental technological advance but signals the emergence of an entirely new creative and economic paradigm—one Siu (2025) compellingly describes as "the end of the knowledge economy" and the birth of a fundamentally altered landscape for human ingenuity. Aneesh Raman, in discussing these changes, observes with wry clarity: "It's only been the past couple decades that work has been about our intellectual abilities" (Siu, 2025). Before that, humans toiled primarily through physical exertion—on farms, in factories, and in manual labor. The knowledge economy thus appears shockingly ephemeral, a transient blip now yielding rapidly to an AI-driven "creativity economy," where distinctly human skills like imagination, adaptability, and social collaboration reign supreme. Raman succinctly captures this shift: In the age of AI, workers must "disrupt yourself or be disrupted"—an imperative equally applicable to creatives, academics, and institutions entrenched in traditional models.

In this evolving "creativity economy," generative AI reshapes traditional roles by magnifying rather than replacing human imagination and collaborative potential. LinkedIn's projections, cited by Siu (2025), underscore that the future belongs to those who embrace a new set of distinctly human skills: ideation, emotional intelligence, adaptability, and the "curatorial mind." The so-called curatorial mind refers to human oversight and the critical capacity to assign credibility, depth, and meaning to generative outputs. These capabilities position humans as indispensable arbiters and interpreters of creative outputs, responsible for distinguishing genuinely innovative material from algorithmic noise—or as one commentator dryly noted, "human curation will separate genuine creative gold from the endless AI-generated slop" (Siu, 2025). Such critical discernment, coupled with adaptive innovation, forms the core competencies of the next era in creativity. Workers, scholars, and artists alike must adopt this mindset, recognizing these tools not as threats, but as invaluable assets amplifying human potential. The blunt truth of this new paradigm is clear, as Raman asserts: The imperative now is starkly straightforward—"disrupt yourself or be disrupted."

1.2 AI, Authorship, and New Norms of Creation

There is little doubt that the latest iteration of a nearly century-old technology has reshaped creative industries more swiftly than any previous technological development, radically expanding the scope and speed of content generation and artistic exploration. Consider ChatGPT, introduced by OpenAI in late 2022, which astoundingly reached one million users in merely five days, escalating to over 100 million monthly active users by January 2023 (Gordon, 2023). Alongside similar platforms like Claude, Stable Diffusion, and Runway, such generative systems have rapidly broadened the creative landscape—transforming how music, visual art, literature, and multimedia content are produced, disseminated, and monetized (Feuerriegel et al., 2024). Unlike conventional creative software such as Photoshop, which rely on direct human manipulation of existing material, these tools employ machine learning (ML) algorithms trained on vast datasets, enabling models to synthesize entirely novel content by recombining learned patterns and elements (Epstein et al., 2023). This generative capacity simulates aspects of human creativity, introducing fundamentally new possibilities—and equally unprecedented legal and ethical questions.

At the core of these controversies is the uncertainty surrounding human involvement in AI-generated content. These systems autonomously interpret user prompts to create original outputs without directly replicating specific data points from their training material (Risi & Togelius, 2020). Such technological capability, often producing results indistinguishable from human-created works, has profound implications for traditional understandings of authorship and copyright protection. Abdikhakimov (2023) notes this ambiguity arises primarily because existing legal frameworks depend heavily on direct, intentional human authorship—criteria difficult to apply when a machine autonomously composes text, imagery, or music. Historically, copyright law has focused squarely on protecting creative outputs clearly traceable to human intent and originality (Ploman & Hamilton, 2024). Thus, current copyright paradigms strain to accommodate this algorithmically driven creativity, challenging traditional definitions of originality and authorship.

1.2.1 The Anxiety of Lost Creative Autonomy

These developments have spurred intense debate over the implications for creative disciplines and society more broadly. For some, including mathematician Marcus du Sautoy (2019), these smart systems promise exciting frontiers, invigorating traditional artforms with new avenues for experimentation and innovation. Conversely, Manyika (2022) highlights profound anxiety, cautioning that rapid technological advancement risks diminishing traditional creative methodologies and blurring the lines of artistic intent. Paul Murphy (2022) further underscores this tension, noting significant pushback from traditionally trained artists and designers toward the accelerating presence of AI-inspired art within mainstream creative culture. These mixed reactions reflect broader societal uncertainty, as creative fields struggle to reconcile evolving technologies with established human-centric frameworks.

The distinctions between generative tools and traditional digital design software highlight the complexities at play. Chavez (2024) contrasts these platforms—where algorithms synthesize images or text from user prompts based on learned styles—with established digital tools such as Photoshop or Adobe Illustrator, which function primarily as digital canvases. These older technologies demand direct, intentional human action and precise user control. By contrast, generative models leverage complex algorithmic decision-making processes, fostering a collaborative dynamic wherein human users guide but do not fully dictate creative outcomes. This nuanced interplay between user prompts and algorithmic generation underscores deeper philosophical questions about authorship, agency, and the evolving human–machine creative partnership.

Even prior to recent generative models, scholars sought conceptual frameworks better suited to understanding the expanding role of technology within creative processes. Messer (2024), for example, argues for reevaluating traditional perceptions of art and creativity by advocating an integrated, collaborative view of human and machine contributions. Likewise, Ng (2021) portrays the interaction between human and machine creativity as complementary rather than adversarial, while Tao (2022) proposes an "actor network" model emphasizing mutual collaboration between creators and algorithms, each capitalizing on the other's strengths. Coeckelbergh (2017) similarly suggests adopting a more "poetic" conception of creativity, one that embraces these tools as a legitimate partner, thus enriching creative possibilities and challenging entrenched artistic conventions. These conceptual shifts have prompted a reconsideration of creativity itself, inviting integration of artistic principles and methodologies within broader academic inquiry. Ahmed (2022), notably, highlights the necessity of incorporating humanistic and emotional elements—such as memory, senses, and subjective experience—into these hybridized creative contexts. For example, Ahmed advocates for treating AI-generated installations and immersive media not simply as finished products, but as integral components of ongoing creative dialogue. Through this lens, emotional and experiential responses elicited by machine-enhanced art themselves constitute meaningful aspects of the creative design process, thus encouraging a holistic, human-centered integration of technology into artistic practice.

Inevitably, such discourse intersects with longstanding debates about creative autonomy and the essential nature of artistic expression. Csikszentmihályi's influential creativity model (Sternberg, 1988), comprising domain knowledge, innovative individuals, and expert evaluators, provides a useful framework for evaluating contributions. In this context, AI serves as a partner within an established field of knowledge, while human experts assess the value and innovation of its outputs. Jennings (2010) similarly defines creative autonomy as requiring intentionality, evaluative independence, and the capacity for generating novel variations. Yet Dreier and Andina (2022) caution that creativity inherently depends upon human judgment; external validation by domain experts ultimately determines whether prompted outputs merit recognition as genuinely creative. This nuanced relationship calls for reframing conventional notions linking creativity strictly to human intention. Warsh (2024) reminds readers that while all art emerges from creative processes, not all creative processes yield works of art; innovation encompasses everything from scientific discoveries to entrepreneurial strategies. The rise of these latest models thus broadens the conceptual boundaries of creativity, pushing beyond traditional confines tied exclusively to human experience. Literary experiments—such as algorithmically generated poetry or interactive storytelling applications—further demonstrate this broader perspective, affirming creativity as a multifaceted and expansive phenomenon.

1.2.2 Brave New World, Breakneck Speed

But beyond literary or poetic use, generative writing has grown exponentially in nearly all areas of content creation, especially in industry. In fact, research increasingly confirms that generative writing tools are being adopted rapidly across industries, irrespective of demands for transparent disclosure. For example, Liang et al. (2025) conducted a comprehensive analysis of AI-assisted writing spanning consumer complaints, corporate communications, job postings, and press releases from international organizations between January 2022 and September 2024. The scale was staggering: Their dataset encompassed over 687,000 consumer complaints, more than 537,000 corporate press releases, approximately 304 million job postings, and nearly 16,000 United Nations (UN) statements. Following ChatGPT's explosive launch in November 2022, generative assistance surged dramatically. By late 2024, large language model (LLM)–generated text accounted for 18% of consumer complaint narratives, 24% of corporate communications, just under 10% of job postings at small enterprises (with higher adoption among younger firms), and roughly 14% of UN press releases. Although adoption has since stabilized, these figures reveal an undeniable reality: Automated text generation is already integral to how organizations communicate, regardless of ethical qualms about disclosure.

This pervasive integration of automated text generation has provoked a profound reevaluation of academic integrity, scholarship, professional norms, and the definition of plagiarism itself. Traditionally, plagiarism depended upon clear acknowledgment of original human authors, typically marked by quotation or substantial paraphrase accompanied by citations (Bouville, 2008). Paraphrasing, beyond mere synonym substitution, demanded substantial reconfiguration of original ideas while preserving their core meaning (Fenton & Gralla, 2020). Yet the seamless, human-like outputs of generative models have muddied these once-straightforward distinctions, challenging institutions to redefine plagiarism in an era where "original thought" can be convincingly simulated by machines (Dehouche, 2021). Traditional definitions built upon explicit human attribution seem increasingly obsolete in the face of automated generative capacities, raising thorny questions about what constitutes original authorship when the boundary between creator and machine grows blurred.

Most anxieties about academic, creative, and scholarly integrity now pivot around large language models (LLMs), like GPT (Generative Pre-trained Transformer), grounded in advanced machine learning and natural language processing methodologies (Wahle et al., 2022). These increasingly sophisticated tools acquire linguistic fluency by ingesting massive datasets covering diverse subject matter and styles, thereby mastering intricate patterns of grammar, syntax, semantics, and even pragmatic contextual nuances (Yu et al., 2023). Transformer architectures permit these models remarkable precision, generating coherent text by accurately capturing long-range dependencies between words (Zaheer et al., 2020). Consequently, LLMs proficiently handle diverse language tasks—text completion, translation, summarization—delivering outputs indistinguishable from human composition in many contexts (Hadi et al., 2023). Such realism has made them invaluable across fields, from drafting complex legal texts to assisting software development, inevitably fueling concerns about distinguishing machine-generated from human-authored content (Jonsson & Tholander, 2022).

These rapid technological advancements significantly complicate established educational and professional practices, particularly within English composition courses, traditionally predicated on writing as reflective of student intellectual engagement (Arapoff, 1967; Jambeck & Winder, 1990; Rahmat, 2020). Writing instruction, fundamentally a tool for developing critical thought and analytical clarity, now contends with generative tools capable of producing sophisticated texts with minimal student input (Roe et al., 2023; Sinaga & Feranie, 2017). This, the argument goes, directly threatens the conventional goal of fostering independent intellectual development. Furthermore, attempts to safeguard academic integrity through detection software, such as Turnitin, have proven problematic. Analysis from May 2023 revealed that among 38.5 million submissions, nearly 10% included substantial (over 20%) AI-generated content, with 3.5% exhibiting between 80% and 100% AI authorship (Chechitelli, 2023). These tools, however, struggle with accuracy—particularly with creative or structured texts like poetry, scripts, or annotated bibliographies—often yielding both false positives and negatives that necessitate manual oversight (Chaka, 2023; Rashidi et al., 2023). False positives are even more pronounced for those with neurodivergent conditions like attention deficit hyperactivity disorder (ADHD) or autism, as well as writers for whom English is their second language.

I.2.3 Accelerators and Gatekeepers

Moreover, generative collaboration between humans and machines (as noted) profoundly unsettles traditional conceptions of intellectual property (IP). As

Oleksy (2023) underscores, joint human–machine authorship creates significant ambiguities around originality and ownership, challenging legal and ethical norms governing plagiarism. These developments demand a fundamental reassessment of how future scholars are trained. Hutson and Plate (2023) suggest educators must thoughtfully integrate generative technologies into curricula in ways that augment—rather than replace—critical analysis and independent thinking, ensuring students retain agency, but, more importantly, gain invaluable skills that will be used in the field. Yet the rapid proliferation of these tools has opened a notable gap between educator-publisher familiarity and student-author usage rates. NerdyNav (2024) data illustrate stark disparities: 82% of university professors were aware of generative tools like ChatGPT, compared to only 55% of K-12 teachers. Among students, generational divides are evident; millennials and Gen Z (13.5%) adopt generative platforms at nearly double the rate of Generation X (7.9%) and Baby Boomers (7.2%), with the Silent Generation trailing at 5.3%. Such generational discrepancies exacerbate skepticism and mistrust among traditional editors and publishers regarding these new technologies (Kreps et al., 2023). This generational gap also underscores the pressing need for increased AI literacy among pedagogues, who must rapidly adapt traditional pedagogical approaches to bridge the widening divide between academic norms and technological reality (Southworth et al., 2023; Su et al., 2023). The same is true for publishers and those involved in the scholarship lifecycle from proposal to reviewer feedback to editing to publication. The challenge is clear: To navigate the inevitable rise of generative tools, academia and industry must evolve.

Despite widespread acknowledgment—and perhaps even grudging enthusiasm—among academics about the necessity of adapting to generative tools, the chorus of resistance remains remarkably vigorous, particularly among traditional gatekeepers. Indeed, many readers now scanning this page may themselves be skeptics who have likely seen enough educational fads to recognize (or suspect) yet another fleeting techno-utopian dream. Within academia, particularly in the humanities and fine arts, professional norms and identities are not merely defended—they are zealously guarded, often by professors whose self-worth is intimately bound to their mastery of a discipline's traditional methods. After all, the established professor training the next generation of English PhDs or Studio Art MFAs serves as the primary arbiter of what counts as authentic scholarly or artistic endeavor. Ironically, these gatekeepers—the very individuals responsible for preparing students to navigate professional landscapes now increasingly dominated by AI—are often the staunchest opponents of its integration. Such resistance invites a provocative question: Are academics actually defending intellectual rigor, or simply hedging against their own professional obsolescence?

Historically, technological innovations have a knack for exposing precisely these existential vulnerabilities in academic identity. Consider the medieval monk who surely viewed Gutenberg's printing press as a demonic contraption, set upon extinguishing the divine artistry of hand-copying manuscripts. Or imagine the indignant professor whose cherished typewriter was swiftly replaced by personal computers that rudely highlighted typos with squiggly red lines. Each technological shift threatened, at least momentarily, to erase the painstaking craftsmanship once synonymous with intellectual labor. Yet each crisis passed, leaving scholars grudgingly reliant upon—and, in time, perhaps affectionate toward—the very technologies they initially despised. The pattern is clear: Resistance to technological disruption is as inevitable as its eventual acceptance.

The anxiety surrounding AI today mirrors past disruptions, revealing less about any true threat to scholarly rigor than about academia's perennial discomfort with change itself. As such, English professors today might recoil at the idea of producing scholarship entirely by hand, with fountain pen and parchment; yet they balk similarly at integrating generative writing tools into their workflows, reflecting deep-rooted beliefs about what constitutes authentic scholarly labor (Messer-Davidow, 2002). Socrates, infamous for decrying the written word as detrimental to memory and genuine intellectual discourse, might well have offered skeptical commentary had he lived to witness generative models autonomously drafting persuasive essays. Indeed, if machine algorithms produce the sentences, whose ideas are ultimately represented—author or algorithm? The crux of scholarly anxiety surrounding LLMs, such as ChatGPT and Claude, hinges precisely upon the perceived erosion of human intellectual engagement. Yet, given historical precedent, this anxiety appears less about technology itself and more about protecting entrenched notions of academic identity and scholarly authenticity.

The reluctance to adopt generative tools into academia extends beyond pragmatic concerns—such as plagiarism—to profoundly existential fears about the changing nature of scholarly identity. Traditionally, scholars have viewed their roles as vocations rather than mere professions, echoing a quasi-religious devotion historically observed in medieval universities like Bologna and Oxford, where scholars embodied an almost priestly dedication to their disciplines (Neumann, 2009; Perkin, 2014). Academia, unlike professions rooted purely in skill or trade, intertwines deeply with personal identity, valorizing the prolonged effort, meticulous attention, and individual mastery required for original research. Faculty members thus often perceive AI assistance as diluting or even undermining the enchanted sense of vocation, replacing human endeavor with algorithmic expedience. Winick (2018) aptly captures this sentiment, noting scholars' fears that automation might erode the personal fulfillment derived from rigorous intellectual craftsmanship and thereby diminish the intrinsic value of their contributions.

Research clearly indicates that generative writing tools are rapidly becoming integral to student workflows, regardless of faculty skepticism. For instance, Liang et al. (2025) reported a dramatic rise in the prevalence of AI-generated text in various domains of professional communication, including educational contexts. The generational divide in the adoption of such tools further underscores this trend, revealing pronounced disparities between younger students and their professors (Kreps et al., 2023). Millennials and Generation Z students embrace generative technologies, like ChatGPT, at rates significantly outpacing their Generation X and Baby Boomer professors, fueling pronounced pedagogical discomfort among educators (Southworth et al., 2023; Su et al., 2023). Such generational gaps intensify academic anxieties, not only over issues of originality and authorship but also around educators' perceived loss of control over student learning processes and outcomes. Indeed, the speed with which these generative tools have entered student and scholarly practices, leaving faculty and publishers scrambling to adjust pedagogies and guidelines that were, until recently, comfortably static, highlights the academy's uneasy relationship with rapid technological shifts.

I.2.4 AI Guilt and the Identity Crisis in Creativity

Yet the integration of AI does not inherently threaten scholarly identity—provided that institutions reconsider what constitutes scholarly contribution. For instance, Ahmed (2022) argues compellingly for integrating generative media into academic discourse not simply as finished artifacts but rather as dynamic catalysts of emotional, experiential, and intellectual engagement. This reconceptualization positions the technology not as a replacement but as a complementary partner, capable of broadening the scope of academic inquiry and facilitating innovative interdisciplinary collaborations. Similarly, Jennings (2010) emphasizes the need for these systems to possess "creative autonomy," suggesting that true integration into creative practices must recognize the algorithmic ability to independently apply and adapt standards.

Addressing entrenched resistance thus requires confronting what scholars now term "AI guilt"—the subconscious conviction that true scholarly value correlates directly to invested effort and time (Gull et al., 2023; Lively & Hutson, 2024). Yet the merit of scholarship has historically depended less upon effort than on outcomes, innovation, and impact. Few would dispute Albert Einstein's (1879–1955) genius simply because he devised revolutionary ideas in his spare time, without prolonged labor. Consequently, academics and scholars might best respond to generative technologies by reassessing entrenched biases that equate scholarly worth with manual intellectual labor, instead valuing the originality and impact of outputs—regardless of the precise methods involved (Oleksy, 2023). It seems likely that success in this emerging era relies upon the skillful blending of human discernment with algorithmic assistance. This shift demands a reconceptualization of academic identity itself. Rather than viewing scholarly value as contingent upon exhaustive individual effort, institutions might embrace collaborative models where human intellect directs, refines, and enhances AI-generated insights. Such an approach retains human judgment and intention as central, while fully exploiting algorithmic efficiencies. Indeed, as Hutson and Plate (2023) recommend, educators must now actively incorporate generative tools into curricula, emphasizing analytical evaluation, curation, and critical discernment as core competencies. By shifting pedagogical focus from mere manual execution to sophisticated critical engagement with algorithmic outputs, academia can preserve scholarly rigor while preparing students for inevitable technological realities.

Therefore, the evolving capabilities of generative technologies, and the existential anxieties they provoke, reveal a familiar historical narrative—one in which scholars and creators routinely greet technological innovation with suspicion, skepticism, and outright resistance. The creative impulse has never comfortably coexisted with transparency demands; nor has authenticity ever depended upon the exhaustive disclosure of methods, despite the persistent contemporary insistence otherwise. From Socrates' anxiety about writing's impact on memory to Duchamp's provocative interrogation of the nature of art itself, creative practices have continually thrived precisely because their origins remain partially concealed, ambiguous, and even mysterious. Contemporary pressures mandating rigorous transparency regarding generative processes thus misunderstand the inherent nature of creativity and authorship, imposing constraints that are both historically unprecedented and fundamentally contrary to artistic impulse. Indeed, such expectations risk imposing artificial—and ultimately untenable—limitations on how creators explore and innovate.

This foregrounding framing has unpacked the uncomfortable reality that contemporary insistence on AI disclosure represents a profound misunderstanding of both historical precedent and the intrinsic nature of creative thought. Younger generations adopting generative tools at record speeds have only intensified traditionalists' anxieties, highlighting a stark generational gap in technological fluency. Academics, those venerable gatekeepers, now face a pivotal choice: stubbornly cling to familiar yet outdated methodologies, or embrace generative collaboration as the next logical evolution in intellectual craftsmanship. Moving forward, this manuscript will demonstrate why comprehensive transparency in AI-assisted creation is not merely impractical, but fundamentally impossible. After all, expecting artists and scholars to document each algorithmic whisper influencing their work is as absurd as demanding Shakespeare reveal precisely which ale inspired Hamlet's existential musings. In the following chapter, we explore these tensions further—examining historical resistance to technological change and the inherent folly of transparency demands in our increasingly algorithmic age.

References

Abdikhakimov, I. (2023, June). Unraveling the copyright conundrum: Exploring AI-generated content and its implications for intellectual property rights. *International Conference on Legal Sciences, 1*(5), 18–32.

Ahmed, D. (2022). Senses, experiences, emotions, memories: Artificial intelligence as a design instead of for a design in contemporary Japan. *Intelligent Buildings International, 14*(2), 133–150.

Arapoff, N. (1967). Writing: A thinking process. *Tesol Quarterly, 1*(2), 33–39.

Boden, M. A. (1978). Artificial intelligence and Piagetian theory. *Synthese, 38*(3), 389–414.

Boden, M. (1994). Creativity and computers. In T. Rickards, M. A. Runco, & S. Moger (Eds.), *Artificial intelligence and creativity: An interdisciplinary approach* (pp. 3–26). Springer.

Bouville, M. (2008). Plagiarism: Words and ideas. *Science and Engineering Ethics, 14*, 311–322.

Britnell, J. (1989). Poetic fury and prophetic fury. *Renaissance Studies, 3*(2), 106–114.

Candy, L. (2002). Introduction: Creativity and cognition — Part I: Perspectives from the third symposium. *Leonardo, 35*(1), 55–57.

Chaka, C. (2023). Detecting AI content in responses generated by ChatGPT, YouChat, and Chatsonic: The case of five AI content detection tools. *Journal of Applied Learning & Teaching, 6*(2), 94–104.

Chavez, C. (2024). *Adobe Photoshop classroom in a book 2025 release.* Adobe Press.

Chechitelli, A. (2023, May 23). AI writing detection update from Turnitin's Chief Product Officer. *Turnitin.* https://www.turnitin.com/blog/ai-writing-detection-update-from-turnitins-chief-product-officer

Coeckelbergh, M. (2017). Can machines create art? *Philosophy & Technology, 30*(3), 285–303.

Danto, A. C. (2013). *What art is.* Yale University Press.

Darshan, G. (2014). The origins of the foundation stories genre in the Hebrew Bible and ancient eastern Mediterranean. *Journal of Biblical Literature, 133*(4), 689–709.

Dehouche, N. (2021). Plagiarism in the age of massive Generative Pre-trained Transformers (GPT-3). *Ethics in Science and Environmental Politics, 21,* 17–23.

Dorter, K. (1973). The Ion: Plato's characterization of art. *The Journal of Aesthetics and Art Criticism, 32*(1), 65–78.

Dreier, T., & Andina, T. (Eds.). (2022). *Digital ethics: The issue of images* (Vol. 11). Hart.

Du Sautoy, M. (2019). The creativity code: Art and innovation in the age of AI. In *The creativity code.* Harvard University Press.

Elton, M. (1995). Artificial creativity: Enculturing computers. *Leonardo, 28*(3), 207–213.

Epstein, Z., Hertzmann, A., Investigators of Human Creativity, Akten, M., Farid, H., Fjeld, J., & Smith, A. (2023). Art and the science of generative AI. *Science, 380*(6650), 1110–1111.

Fenton, A. L., & Gralla, C. (2020). Student plagiarism in higher education: A typology and remedial framework for a globalized era. In B. Montoneri (Ed.), *Academic misconduct and plagiarism* (p. 109). Lexington.

Feuerriegel, S., Hartmann, J., Janiesch, C., & Zschech, P. (2024). Generative AI. *Business & Information Systems Engineering, 66*(1), 111–126.

Foster, H. (2014). Exquisite corpses. In L. Taylor (Ed.), *Visualizing theory* (pp. 159–172). Routledge.

Goldsmith, S. (1983). The readymades of Marcel Duchamp: The ambiguities of an aesthetic revolution. *The Journal of Aesthetics and Art Criticism, 42*(2), 197–208.

Gordon, C. (2023, February 2) ChatGPT is the fastest growing app in the history of web applications. *Forbes.* https://www.forbes.com/sites/cindygordon/2023/02/02/chatgpt-is-the-fastest-growing-ap-in-the-history-of-web-applications/

Growcoot, M. (2025, February 12). This is the first-ever AI image to be granted copyright protection. *Peta Pixel.* https://petapixel.com/2025/02/12/this-is-the-first-ever-ai-image-to-be-granted-copyright-protection-a-slice-of-american-cheese/

Grundberg, A. (2021). *How photography became contemporary art: Inside an artistic revolution from pop to the digital age.* Yale University Press.

Gull, A., Dilawar, S., & Sher, F. (2023). Data-driven Artificial Intelligence at the crossroads: Investigating the role of affective job insecurity in the relationship between Artificial Intelligence identity threat and employee well-being. *The Asian Bulletin of Big Data Management, 3*(1), 18–34.

Hadi, M. U., Qureshi, R., Shah, A., Irfan, M., Zafar, A., Shaikh, M. B., Akhtar, N., Hassan, S. Z., Shoman, M., Wu, J., Mirjalili, S., & Shah, M. (2023). A survey on large language models: Applications, challenges, limitations, and practical usage. *Authorea Preprints.*

Hamilton, A. (2024). *Art and entertainment: A philosophical exploration.* Taylor & Francis.

Hutson, J., & Plate, D. (2023). *Human-AI collaboration for smart education: Reframing applied learning to support metacognition.* IntechOpen.

Jambeck, K. K., & Winder, B. D. (1990). Vygotsky, Werner, and English composition: Paradigms for thinking and writing. *Writing on the Edge, 1*(2), 68–79.

James, W., & Dewey, J. (2005). *James and Dewey on belief and experience.* University of Illinois Press.

Jennings, K. E. (2010). Developing creativity: Artificial barriers in artificial intelligence. *Minds and Machines, 20,* 489–501.

Jiménez, J., Ades, D., Sebbag, G., & Thyssen-Bornemisza Museum (Madrid). (2013). *Surrealism and the dream* (p. 50). Museo Thyssen-Bornemisza.

Jonsson, M., & Tholander, J. (2022, June 20–23). Cracking the code: Co-coding with AI in creative programming education. In *Proceedings of the 14th conference on creativity and cognition* (pp. 5–14). Venice, Italy.

Klukosky, F. P., & Kohel, M. D. (2024). An update on the state of play with generative Artificial Intelligence and intellectual property issues. *Intellectual Property Litigation, 34*(1), 10–17.

Kreps, S., George, J., Lushenko, P., & Rao, A. (2023). Exploring the artificial intelligence "Trust paradox": Evidence from a survey experiment in the United States. *PLoS One, 18*(7), e0288109.

Lang, K. (Ed.). (2019). *Field notes on the visual arts: Seventy-five short essays.* Intellect Books.

Liang, W., Zhang, Y., Codreanu, M., Wang, J., Cao, H., & Zou, J. (2025). The widespread adoption of large language model-assisted writing across society. *arXiv preprint arXiv:2502.09747.*

Lively, J., & Hutson, J. (2024). The role of student motivation in integrating AI into web design education: A longitudinal study. In *Forum for Education Studies, 2*(2), 1242.

Manu, A. (2024). *Transcending imagination: Artificial intelligence and the future of creativity.* CRC Press.

Manyika, J. (2022). Getting AI right: Introductory notes on AI & society. *Daedalus, 151*(2), 5–27.

Martiniano, C. (2016). The scientization of creativity: "Innovate or die!" *Journal of the Midwest Modern Language Association, 49*(2), 161–190.

Messer, U. (2024). Co-creating art with generative artificial intelligence: Implications for artworks and artists. *Computers in Human Behavior: Artificial Humans, 2*(1), 100056.

Messer-Davidow, E. (2002). *Disciplining feminism: From social activism to academic discourse*. Duke University Press.

Mitter, P. (2008). Decentering modernism: Art history and avant-garde art from the periphery. *The Art Bulletin, 90*(4), 531–548.

NerdyNav. (2023, January 5). ChatGPT cheating statistics & impact on education (2024). https://nerdynav.com/chatgpt-cheating-statistics/

Neumann, A. (2009). *Professing to learn: Creating tenured lives and careers in the American research university*. JHU Press.

Ng, J. (2021). An alternative rationalisation of creative AI by de-familiarising creativity: Towards an intelligibility of its own terms. In P. Verdegem (Ed.), *AI for everyone?: Critical perspectives* (pp. 49–66). University of Westminster Press.

Oleksy, E. (2023). That thing ain't human: The artificiality of "human authorship" and the intelligence in expanding copyright authorship to fully-autonomous AI. *Cleveland State Law Review, 72*(1), 263.

Paul Murphy, B. (2022, December 15). Is Lensa AI stealing from human art? An expert explains the controversy. *Science Alert*. https://www.sciencealert.com/is-lensa-ai-stealing-from-human-art-an-expert-explains-the-controversy

Perkin, H. (2014). History of universities. In P. Altbach (Ed.), *International higher education volume 1* (pp. 169–204). Routledge.

Ploman, E. W., & Hamilton, L. C. (2024). *Copyright: Intellectual property in the information age*. Taylor & Francis.

Racter. (1984). *The policeman's beard is half constructed*. Warner Books.

Rahmat, N. H. (2020). Thinking about thinking in writing. *European Journal of Literature, Language and Linguistics Studies, 3*(4), 20–37.

Rashidi, H. H., Fennell, B. D., Albahra, S., Hu, B., & Gorbett, T. (2023). The ChatGPT conundrum: Human-generated scientific manuscripts misidentified as AI creations by AI text detection tool. *Journal of Pathology Informatics, 14*, 100342.

Risi, S., & Togelius, J. (2020). Increasing generality in machine learning through procedural content generation. *Nature Machine Intelligence, 2*(8), 428–436.

Roe, J., Renandya, W. A., & Jacobs, G. M. (2023). A review of AI-powered writing tools and their implications for academic integrity in the language classroom. *Journal of English and Applied Linguistics*, *2*(1), 3.

Rudinow, J. (1981). Duchamp's Mischief. *Critical Inquiry*, *7*(4), 747–760.

Sinaga, P., & Feranie, S. (2017). Enhancing critical thinking skills and writing skills through the variation in non-traditional writing task. *International Journal of Instruction*, *10*(2), 69–84.

Siu, E. (2025, February 26). "The knowledge economy is on the way out." These are the skills workers will need in the age of AI, says LinkedIn. *CNBC*. https://www.cnbc.com/amp/2025/02/26/the-skill-humans-can-leverage-as-ai-disrupts-workforces-globally.html

Smith, P. (2013). Cézanne's "primitive" perspective, or the "view from everywhere." *The Art Bulletin*, *95*(1), 102–119.

Somaya, D., & Varshney, L. R. (2020). Ownership dilemmas in an age of creative machines. *Issues in Science and Technology*, *36*(2), 79–85.

Southworth, J., Migliaccio, K., Glover, J., Reed, D., McCarty, C., Brendemuhl, J., & Thomas, A. (2023). Developing a model for AI across the curriculum: Transforming the higher education landscape via innovation in AI literacy. *Computers and Education: Artificial Intelligence*, *4*, 100127.

Stankiewicz, M. A. (2000). Discipline and the future of art education. *Studies in Art Education*, *41*(4), 301–313.

Sternberg, R. J. (Ed.). (1988). *The nature of creativity: Contemporary psychological perspectives*. CUP Archive.

Su, J., Ng, D. T. K., & Chu, S. K. W. (2023). Artificial intelligence (AI) literacy in early childhood education: The challenges and opportunities. *Computers and Education: Artificial Intelligence*, *4*, 100124.

Tao, F. (2022). A new harmonisation of art and technology: Philosophic interpretations of artificial intelligence art. *Critical Arts*, *36*(1–2), 110–125.

Taylor, R. P., Micolich, A. P., & Jonas, D. (2002). The construction of Jackson Pollock's fractal drip paintings. *Leonardo*, *35*(2), 203–207.

Wahle, J. P., Ruas, T., Kirstein, F., & Gipp, B. (2022). How large language models are transforming machine-paraphrased plagiarism. *arXiv preprint arXiv:2210.03568.*

Warsh, L. (Ed.). (2024). *JR-isms*. Princeton University Press.

Whitford, W. G. (1923). Brief history of art education in the United States. *The Elementary School Journal, 24*(2), 109–115.

Winick, M. (2018). Scholarly enchantment. *Nineteenth-Century Literature, 73*(2), 187–226.

Yu, D., Li, L., Su, H., & Fuoli, M. (2023). Assessing the potential of LLM-assisted annotation for corpus-based pragmatics and discourse analysis: The case of apology. *International Journal of Corpus Linguistics, 29*(4), 534–561.

Zaheer, M., Guruganesh, G., Dubey, K. A., Ainslie, J., Alberti, C., Ontanon, S., Pham, P., Ravnla, A., Wang, Q., Yang, L., & Ahmed, A. (2020). Big bird: Transformers for longer sequences. *Advances in Neural Information Processing Systems, 33*, 17283–17297.

CHAPTER 1

The Illusion of Transparency

Transparency is a lovely concept—comforting, virtuous-sounding, and utterly unattainable. This chapter dismantles this well-intentioned myth. The call for complete disclosure in creative processes is not merely impractical but fundamentally at odds with the chaotic reality of creativity itself. From the amusing futility of documenting every neuron fired during creative epiphanies, to the outright absurdity of tracing algorithmic whispers within generative technologies, the notion that creators can—or should—fully disclose their methods is a modern illusion devoid of historical precedent. Diving into past anxieties, this chapter exposes transparency as academia's latest doomed stand against inevitable change. Therefore, demanding exhaustive disclosure about how art and scholarship come to be is akin to politely asking a magician to explain the rabbit:

1.1 The Impossibility of Complete Disclosure

When photography burst onto the scene in the mid-nineteenth century, painters collectively gasped—not simply because a mechanical device could capture reality with unsettling accuracy but because the very nature of their artistry appeared under threat. Suddenly, brushstrokes carefully perfected over years of training seemed redundant. But imagine for a moment if these same anxious painters had been forced to meticulously document precisely how they used their new rival technology. How many exposures were required? Which precise settings yielded the ideal shadows? Which moment, exactly, did inspiration strike—perhaps between sips of coffee or in a momentary daydream? Such demands sound comically absurd today, yet proponents of transparency around generative tools in contemporary creative work earnestly insist upon just such exhaustive documentation. They argue, sometimes with indignation, that disclosing how, when, and where generative technologies are employed preserves the integrity of authorship, ensures fair attribution, and protects against plagiarism and deception. After all, if an algorithm significantly shapes the final product, should audiences not know the exact nature and extent of its involvement?

Proponents frequently assert that transparency is ethically imperative because it clearly delineates human creativity from automated outputs, reinforcing traditional notions of originality and authorship. This belief stems from an understandable desire for clarity in a rapidly evolving digital landscape where the human touch risks becoming invisible, overshadowed by algorithmic prowess. For instance, educators insist students reveal how much assistance they have received from tools like ChatGPT or Claude to fairly assess their analytical and creative abilities. Similarly, artists who labor over traditional methods worry their painstaking efforts will be undervalued if generative tools can silently replicate or even surpass their results. Transparency thus promises a comforting illusion: that clear distinctions remain possible between human ingenuity and machine-assisted creations, and that human creative effort can be accurately measured, valued, and protected in an age of automated assistance.

Yet even the simplest technological interventions in creative work have never lent themselves to such rigorous documentation—nor should they. Consider an author writing on a modern laptop equipped with spellcheck software. Could the author realistically document precisely how each sentence came into being, specifying when the software suggested an alternative phrasing or flagged potential errors? Likewise, an artist editing a photograph in Photoshop would find it impossible—and absurdly tedious—to itemize each brush stroke, slider adjustment, or minor enhancement. Would a novelist using Google to verify historical facts carefully document each search query and the exact influence each retrieved page had on the narrative? Demanding such disclosure not only imposes an impractical administrative burden but fundamentally misunderstands the spontaneous, iterative, and often subconscious nature of creativity itself. Indeed, part of creativity's charm—and effectiveness—lies precisely in its unpredictability, ambiguity, and resistance to documentation.

1.1.1 Confessing the Obvious: The Irony of Mandating Disclosure

In essence, the current insistence on transparency for generative technologies over-looks the complex, often intangible realities of the creative process. Artists, writers, and scholars rarely proceed linearly, consciously accounting for every influence and intervention along the way. Ideas percolate beneath the surface, shaped by countless forgotten inputs—conversations overheard, books half-remembered, accidental keystrokes—and now, yes, algorithmic suggestions. Creativity, after all, thrives precisely because its origins remain elusive and opaque. The uncomfortable truth, then, is that transparency in artificial intelligence (AI)–assisted creation, far from being ethically essential or practically achievable, is merely a reassuring fiction—an illusion comforting enough to insist upon, yet impossible enough never to fulfill.

Yet, paradoxically—and perhaps amusingly—the U.S. Copyright Office (2025) demands precisely such exhaustive documentation, despite its inherent impracticality and absurdity. In a turn Kafka himself might appreciate, current guidelines require creators using generative technologies to explicitly record and disclose every incremental intervention made in the process of creation. Consider the groundbreaking yet illustrative case of *A Slice of American Cheese* by Kent Keirsey (Growcoot, 2025): The Copyright Office granted protection only after Keirsey painstakingly documented no fewer than thirty specific creative interventions. Each intervention was meticulously recorded, capturing every iterative prompt given to the algorithm—every tweak, every decision, every directive issued. Ironically, Keirsey's role was predominantly instructive, limited primarily to issuing commands to the algorithm. Thus, the first instance of AI copyright to date required documenting not tangible artistic skill or direct craftsmanship but the act of guiding machine behavior itself. Such stringent requirements illuminate the profound tension between institutional demands for transparency and the inherently opaque, often intuitive nature of creativity—a paradox explored in depth in the next chapter on intellectual property and the precarious status of algorithmically assisted authorship.

Such irony becomes even more apparent when considering the active struggle required to not to utilize generative tools. The pervasive integration of automated assistance into everyday platforms—Instagram, Facebook, TikTok, and even commonplace office software—means that authors and creatives must increasingly exert conscious effort to avoid algorithmic collaboration. Numerous add-on tools, such as the Grammarly app, automatically activate within digital platforms to offer suggestions or, increasingly, to autonomously generate content for the user. Who has not experienced replying to an email only to discover Microsoft Outlook has already helpfully composed readymade responses—just waiting for a click, no further thought required? Indeed, when Microsoft triumphantly announced the incorporation of Copilot into Word, PowerPoint, and their entire software suite November 2023, several colleagues visibly recoiled in existential dread, exclaiming in horror, "Why would they do that? It's awful!" Yet this panicked response underscores a deeper cultural shift already underway: The challenge today is not learning how to use generative technologies but actively resisting their ubiquity.

This phenomenon contributes significantly to what Markowitz (2025) aptly describes as an "existential crisis" in knowledge-based professions. To remain competitive, augmenting oneself with generative tools has rapidly moved from luxury to necessity. OpenAI's Deep Research functions and Grok 3 capabilities already exceed human performance in various analytical tasks—so much so that David Perell (@david_perell), "The Writing Guy," renowned for his popular writing courses, posted on Twitter/X on February 24, 2025 that he had decided to close down his business, openly attributing his decision to large language models' (LLMs) reshaping of the writing profession. With characteristic bluntness, Perell noted: "The world of non-fiction writing has fundamentally changed.... Many of the skills I've built my career on are becoming increasingly irrelevant." He candidly acknowledged that the expertise required to surpass a well-prompted generative model—like ChatGPT's Deep Research reports—is escalating rapidly, outpacing the average writer's capacity. Perell offers a simple heuristic for understanding which forms of writing will endure: Content firmly rooted in personal experience, such as memoirs and biographies, remains resilient precisely because human lives provide nuanced data inaccessible to algorithms. Ironically, as generative tools automate background research with astonishing competence, human authorship increasingly finds value in the very subjectivity once derided as anecdotal or unscientific—first-hand experiences, unique perspectives, and hard-won personal convictions.

1.1.2 Absent Presence: Echoes of Creative Ghosts Past

Yet beneath the fervent debate surrounding synthetic creations lies a deeper philosophical unease: the supposed "soullessness" attributed to works produced by algorithmic means. Humanist critics frequently express this concern, lamenting that generative literature or imagery lacks emotional authenticity, dismissing such outputs as inherently impersonal or detached from human experience. This critique, however, conveniently overlooks a subtler yet equally powerful phenomenon known as "Absent Presence" (Sahay, 1997)—a concept exemplified in numerous literary and artistic works, wherein meaning is profoundly shaped precisely by what or who is not visibly present. Consider Shakespeare's *The Winter's Tale*, in which Queen Hermione, believed dead, exerts a profound influence over the narrative, her absent figure looming large in every line and scene. Characters feel her presence palpably, her impact on their choices and emotional arcs remaining potent despite her apparent demise. Likewise, van Gogh's famous paintings of empty chairs (Figure 1.1), though devoid of human sitters, are profoundly expressive precisely because they evoke the absent individuals, their personalities, and their lingering traces within the silent, empty spaces. The absence itself conveys a narrative; the empty chair becomes more hauntingly human precisely because no person is there.

Figure 1.1: Vincent van Gogh, *Chair*, 1888 (CC O)

This phenomenon is also vividly rendered through Sidney's *Astrophil and Stella*, where Astrophil's constant longing underscores Stella's absence as her very absence heightens emotional intensity, amplifying her power and influence over his creative imagination. Sidney's deliberate paradoxes and yearning metaphors underscore how absent presence shapes creative consciousness and emotional authenticity. Similarly, Marcel Duchamp's *Fountain* (1917) (Figure 4) presents the mundane, industrial urinal as a work of art—not by its physical presence alone but through the provocative absence of traditional craftsmanship and artistic intent. Duchamp's conceptual gambit derives its potency precisely because it rejects conventional forms of visible artistic labor, inviting viewers to grapple with the absent expectations and the traditional notions of authorship and technique typically assumed essential to artistic legitimacy. The artifact's potency rests entirely on the invisible intentions behind its selection rather than its visible form.

Generative tools, including platforms like ChatGPT and DALL-E, invoke precisely this kind of spectral human influence. Although algorithmically composed outputs might lack obvious direct human labor, they bear the subtle but undeniable imprint of human creativity, culture, and collective intellectual history. Like Shakespeare's spectral Hermione or van Gogh's vacant chair, these algorithmic works carry within their fabric traces of their human origins—millions of lines of poetry, prose, images, and art that informed the model's internal logic. The vast training datasets upon which these models rely, composed entirely of human-authored texts and images, effectively encode an expansive yet invisible network of human thought, creativity, and culture into each generated output. This layering of human influence becomes evident, if invisible, in every generated image or sentence. Just as an author's voice reverberates with echoes of literary predecessors—Dickens' humanity, Austen's wit, Woolf's introspection—so too do generative outputs silently reference a collective human consciousness. Dismissing algorithmically generated works as "soulless" ignores the profound human resonance encoded in their construction.

The irony embedded in critiques of algorithmic creation thus emerges clearly: Those dismissing generative outputs as sterile or "inhuman" seemingly overlook the invisible threads of human consciousness and culture embedded in every pixel and phrase. Equally ironic is the fact that human creators themselves rarely recognize or explicitly document the myriad subtle influences shaping their work—ideas gleaned from forgotten conversations, subconscious memories, or past readings. The generative algorithm simply makes this hidden process more explicitly visible, prompting critics to demand impossible levels of disclosure regarding the origins and influences behind every creative act. But if human creative processes, historically celebrated for their ineffable and intuitive character, evade detailed scrutiny, why should generative works be judged by different standards? In the end, generative creations challenge us not to dismiss them as inherently impersonal or artificial but to recognize them as an image of the nature of human imagination. One might view the output of an LLM as a kind of *reductio ad absurdum* of certain ideas of human originality, but one cannot simply reject the LLM as inhuman.

1.1.3 Creativity by Checklist: Resisting Procedural Absurdity

Yet, for argument's sake, let us temporarily suspend disbelief and entertain the current insistence on exhaustive disclosure of each minuscule step in the creative process. Imagine applying such rigorous transparency to the workflow of a traditional academic scholar—one who prides themself on meticulousness, intellectual integrity, and impeccable citation style. They begin, perhaps innocently enough, with a modest conference abstract. Following its acceptance, this abstract is laboriously expanded into a full presentation, delivered to peers at an academic conference. These colleagues—friendly rivals at best—generously provide feedback (sometimes unrequested, always lengthy), prompting further revisions. This conference paper, now battle-hardened, becomes a journal submission. The journal editor then solicits anonymous reviews, a process guaranteed to generate copious additional suggestions, extensive revisions, and an inevitable crisis of professional confidence. Each suggestion sparks another spiral of research: countless hours at university libraries, carefully combing through analog manuscripts and texts; digitally sifting through JSTOR, ProQuest, and Google Scholar; annotating margins, bookmarking PDFs, and dog-earing pages of physical volumes stacked precariously on desks. Next, collating quotations and citations meticulously recorded in Zotero or EndNote becomes an obsessive preoccupation. Finally, article publication itself draws another round of peer criticism, further adjustments, and thus begins the torturous evolution from a single article into a full-length scholarly monograph. Draft chapters accumulate, each draft subjected to rounds of merciless editing—fact-checking every citation, confirming accuracy against original sources, and proofreading at levels painstaking enough to inspire monastic awe.

Now, let us pause: Imagine a publisher requiring documentation of these steps. Would a scholar faithfully document every conversation at every coffee break, every serendipitous recommendation from a colleague, or each insightful but ultimately forgotten lecture they attended as an undergraduate? Would the scholar dutifully catalog every digital search query entered into Google Scholar or JSTOR, every algorithmic citation suggestion recommended by online databases, or every spell-checked, grammar-corrected sentence in Microsoft Word? The prospect is both comic and horrifying—more an act of obsessive archival performance art than genuine scholarship. Indeed, insisting on transparency at this absurdly granular level turns the very idea of creativity into a parody of bureaucratic box-checking, divorced entirely from intellectual exploration or meaningful contribution to the field.

Now, envision applying similar disclosure requirements when AI is incorporated thoughtfully into this complex academic workflow. Suppose a forward-thinking publisher (variations on this scenario are explored thoroughly in Chapter 5) demands detailed disclosure of precisely how generative tools were utilized. The resulting account might read like a twisted parody:

> The author initially prompted ChatGPT-4 at 8:02 a.m. on February 5, requesting a concise summary of theoretical frameworks related to "academic identity." The prompt "Summarize key concepts of scholarly identity formation from Billot (2010), Kogan (2000), and Neumann (2009) in 200 words or fewer" yielded a rough synthesis, heavily revised and expanded by the author. On February 9 at 11:47 p.m., Claude provided an alternate phrasing for the section on Duchamp's Fountain, though only partially accepted. Three days later, at approximately 2:32 a.m. (after copious coffee consumption), the author prompted ChatGPT, asking: "List ten recent peer-reviewed articles addressing AI integration in higher education, particularly with respect to humanities disciplines." The provided references became a preliminary bibliography—subject, of course, to manual verification, supplementation, and correction through manual database searches. For Chapter 3, illustrative examples were brainstormed using DALL-E 3 at various points between March 4 and April 11, though the author ultimately discarded four out of the five generated concepts as overly surreal, entirely too whimsical, or frankly absurd. On May 5, Grammarly (perhaps the least threatening AI collaborator thus far) suggested twenty-seven minor grammatical corrections, seventeen of which the author accepted grudgingly. Lastly, ChatGPT refined five transitional paragraphs in Chapter 6 on the topic of "absent presence," leading to final revisions reviewed exhaustively by human colleagues at three separate meetings on June 14, July 3, and August 27—each featuring several intense, yet ultimately friendly, scholarly disagreements.

Absurd, certainly; cumbersome, unquestionably. Yet this exhaustive hypothetical disclosure reveals something critical: Demands for absolute transparency become both comical and profoundly impractical when taken to their logical extreme. Far from clarifying authorship or ensuring accountability, such meticulous documentation reduces creative scholarship to a series of ridiculous checklists, an administrative chore serving neither authenticity nor intellectual rigor. The demand for perfect transparency, in short, illuminates only one thing clearly: a fundamental misunderstanding of the nature of creativity itself.

Beyond the unwieldy—and frankly comedic—nature of documenting every creative step, a fundamental aspect eludes capture entirely: the inner workings of the writer's mind. Demands for detailed transparency grossly underestimate the richly tangled web of intellectual, emotional, and experiential threads woven invisibly through each scholarly sentence. An author does not operate merely as a mechanistic scribe meticulously transcribing research notes and citations; rather, the writing process unfolds as an unpredictable interplay between conscious intent and unconscious influence. Scholars routinely synthesize fragments of forgotten articles skimmed as hurried undergraduates, fleeting impressions from half-remembered poetry recited by a grandmother, and echoes from childhood books half-understood yet deeply felt. Methodologically documenting how these subconscious inspirations shape argumentation or rhetoric is not simply difficult—it is impossible.

Further complicating transparency demands is that the act of scholarly writing, like all genuine creativity, is iterative, layered, and nonlinear. Writers pause mid-sentence, distracted momentarily by memories sparked by phrases they themselves just typed; they delete paragraphs, restructure arguments, and suddenly discover, as if miraculously, that their central thesis emerged not from rigorous outlining but rather from an overheard conversation in graduate school—or perhaps from a song lyric heard at random decades before. Such ephemeral mental associations, spontaneous insights, and internal dialogues cannot be itemized, categorized, or documented in any meaningful sense. Yet they remain essential. They constitute the very core of what transforms words on a page into meaningful scholarship. The insistence on transparency and the precise tracing of technological influence overlooks that creative authorship thrives precisely because it draws freely and invisibly from such unpredictable mental intersections and subconscious inspirations.

Therefore, exhaustive documentation misses the point: Writing is a complex dance of analysis, intuition, memory, and imagination—utterly resistant to reductionist breakdowns of procedural steps, technological interventions, and algorithmic influences. Expecting detailed disclosure of how, exactly, a generative model was used disregards the inherently opaque and spontaneous character of the creative intellect itself. Ironically, proponents who argue passionately for transparency in authorship typically cherish literature precisely because it resists straightforward explication. To demand otherwise from today's scholars, writers, and artists is to misunderstand both historical precedent and the intrinsic, glorious chaos of creativity.

1.1.4 Judge the Work, Not the Workshop

The primary contention advanced here—and central to the argument of this book—is that creativity has never been judged by the meticulous documentation of methods or processes but, rather, by the power, impact, and resonance of the finished product itself. Society has long measured artists, writers, and scholars by their outputs, not by how many painstaking steps, revisions, or drafts preceded these outcomes. A reading of Flaubert's accounts of producing only a few words in an evening's work is interesting for a scholar of *Madame Bovary*, but it has no relevance to judging the value of the novel next to Robert Louis Stevenson's *The Strange Case of Dr. Jekyll and Mr. Hyde*. Even if the latter did emerge in one piece in the work of a night's dreaming, this fact of literary history is enjoyable trivia rather than evidence of some calculating sort.

Indeed, the intrinsic value of art, literature, and scholarship has always been grounded in the audience's experience of the final work—not in a detailed audit trail of its creation. Readers cherish Jane Austen's novels not because of a documented trail of her editing decisions but because of the timeless wit and incisive social commentary her stories deliver. The beauty of Michelangelo's Sistine Chapel ceiling is appreciated without requiring documentation of every brushstroke and preparatory sketch, and Beethoven's symphonies are revered without exhaustive accounting of compositional revisions or external influences. Scholarly research similarly stands or falls based on its originality, relevance, and methodological rigor—qualities assessed by peers and readers through the final written argument, not the author's painstaking notes, discarded hypotheses, or citation corrections.

Therefore, contemporary demands for exhaustive disclosure of generative technological involvement betray a profound misunderstanding of what makes any creative or intellectual work meaningful. Audiences—be they readers, viewers, listeners, or scholarly communities—have historically judged creative contributions on their inherent merit, emotional impact, and intellectual rigor, not on the minute details of their production. The introduction of generative tools does not fundamentally alter this reality; what ultimately matters is whether the resulting content engages, inspires, moves, or advances knowledge. Whether achieved through meticulous solitary effort, subconscious inspiration, or algorithmic assistance, it is the quality, resonance, and originality of the finished work itself that will—and indeed should—continue to be the sole criterion upon which creative endeavors are evaluated.

Despite these logical appeals against hegemonic bureaucracy and to creative autonomy, many readers will still remain unmoved—clinging steadfastly to entrenched beliefs about originality, authorship, and the sanctity of human creativity. But beneath these arguments lies a deeper psychological current: the primal unease that accompanies any technology capable of performing distinctly human tasks. Since antiquity, humans have gazed warily at inventions—from automata of ancient Greece to eighteenth-century mechanical chess players—that mimicked human behavior a bit too closely for comfort. This enduring anxiety surfaces vividly whenever technological advancements blur lines previously thought impermeable, unsettling cultural notions of humanity's unique capabilities. Thus, the resistance to generative tools in creative and scholarly practice reflects not merely concerns over transparency or authorship but an ancient, persistent fear—one that reveals far more about us than about the technologies we create.

1.2 Historical Resistance to Technological Change

Historically, resistance to technological innovation often stems less from its practical implications and more from deeply ingrained cultural anxieties—a theme vividly manifesting in contemporary attitudes toward generative tools in scholarly and creative fields. Media portrayals frequently amplify such concerns, painting AI technologies as mysterious "black boxes" whose opaque processes threaten autonomy in scholarly decisions, creative integrity, and intellectual authenticity (Lake, 2024). This apprehension isn't wholly irrational: A scholarly article or literary manuscript, traditionally attributed to a singular human intellect, is now produced by a process invisible to both creator and audience. Yet, what critics decry as unprecedented opacity is precisely the hallmark of all revolutionary technologies. After all, Gutenberg never explained the inner mechanics of his press to readers; nor did Renaissance artists clarify precisely how camera obscura guided their brushstrokes—transparency, evidently, wasn't their strong suit.

Indeed, skepticism toward technological upheaval echoes past cycles of innovation. Cinema initially drew ridicule from theater traditionalists; digital photography faced derision from film purists; and contemporary AI-generated texts and images similarly provoke existential unease. Critics worry openly about the displacement of human roles, framing advanced generative models as existential threats rather than opportunities (Nguyen & Hekman, 2024; Owsley & Greenwood, 2024). Yet, beneath these objections is an ironic undercurrent: Such tools amplify human creativity precisely because they are reflections—often distorted yet undeniably human—of our collective cultural knowledge. The true discomfort perhaps lies not in artificial intelligence but in the mirror it holds up to our own understanding of creativity, authorship, and human exceptionality.

Central to this discomfort is the notion of algorithmic imperfection, with critics particularly fixated on the so-called AI hallucinations, those spontaneous fabrications from LLMs like GPT-4 (Yadlin & Marciano, 2024). Admittedly, early models like ChatGPT 3.5 sometimes resembled inventive undergraduates, creatively fabricating facts and citations alike (Koçak et al., 2025). Yet such missteps are hardly unprecedented—human scholars regularly encounter mis-quoted texts, misunderstood methodologies, and peer-reviewed inaccuracies. The response isn't wholesale rejection but correction and critique; the process, in short, mirrors traditional scholarly rigor. Moreover, contemporary generative models like ChatGPT o1-Pro and the new o3 series already outperform humans in sheer creative volume and novelty, even in controlled comparisons with experienced researchers (Bohren et al., 2024). Ironically, humans augmented by AI sometimes underperform compared to algorithmic creations alone—perhaps reflecting hu-man anxieties more than technical limitations (Bankins et al., 2024; Ramirez & Esparrell, 2024). Thus, generative algorithms can no longer be easily dismissed as "soulless," lacking human nuance or ingenuity; rather, they represent an un-comfortable challenge to traditional conceptions of human creative supremacy.

However, resistance to these generative tools in academia and creative pro-fessions persists—not primarily because of technical inadequacy but due to deeper cultural fears rooted in identity and prejudice. Studies indicate clear biases: AI-generated work is routinely undervalued by human evaluators, even when objectively superior (Hattori et al., 2024). Such biases reflect not objec-tive assessment, but a broader cultural unease, echoing historical technophobic fears. From Luddites smashing industrial looms to contemporary fears of a "Roboapocalypse" (Korać, 2024), the cultural narrative remains consistent: Each technological breakthrough initially sparks existential dread. Yet history repeatedly shows us these fears give way, and technologies integrate seamlessly into creative practice. This resistance must be reframed as a cultural preju- dice—akin to other forms of social discrimination rooted in anxiety about loss of power, control, and status (Rehman et al., 2024). Overcoming this prejudice is imperative, not simply for the practical integration of these powerful tools into scholarly and creative practice but, instead, for realizing their broader potential for social good. After all, one cannot help but wryly observe: Shakespeare, had he worked today, would surely have embraced generative assistants—so long as they didn't demand co-authorship.

1.2.1 Ancient Fears and Modern Anxieties About Machine Autonomy

A substantial body of historical and cultural scholarship demonstrates that resistance to generative technologies arises from deep-seated anxieties about machines encroaching upon distinctly human domains, intertwining these fears with symbolic narratives that date back millennia. Classical myths are rich with mechanical figures and artificial life-forms whose presence challenged the authority of humans, while contemporary science fiction intensifies these themes through more immediate and relatable tropes (Huyssen, 1981; Paschalis, 2015). At the heart of these stories lies an enduring tension: fascination with technological advancement balanced precariously against dread that such tools might someday slip human grasp. Modern portrayals of automated intelligence in mass media frequently exaggerate dystopian possibilities—job displacement, unchecked surveillance, and existential crises of identity—often blending transformative innovation with the specter of irreversible cultural or societal loss. As a result, AI anxieties have seeped into contemporary policy debates, educational strategies, and corporate decision-making, illustrating how deeply embedded cultural frame- works continue to shape the collective interpretation of technological progress. Yet, ironically, the very humans who fret most fervently about the dangers of intelligent machines readily surrender their secrets to the nearest social media app, blissfully unaware of the algorithmic scrutiny they have embraced willingly.

Consider the now-iconic film, *The Terminator* (1984), which introduced audiences to Skynet—an artificial neural network originally designed to streamline military decision-making by removing human error. Skynet's self-awareness, dramatized into catastrophic consequence on August 29, 1997, led to the nuclear oblitera- tion of humanity's dominance on Earth. Judgment Day may never have arrived, thankfully, but the cultural anxiety persists: Fear of AI-driven replacement has permeated public consciousness, even before recent breakthroughs in generative technologies. Indeed, as Aoun (2017) pointedly remarks in *Robot-Proof*, contemporary Americans fear displacement by automation perhaps more keenly than death itself—an impressive existential dread considering our talent for inventing worse fates. Thus, though humanity narrowly escaped Skynet's apocalypse (at least for now), cultural narratives persistently evoke a dystopian trope, blurring the lines between fiction and public perception regarding the real capabilities and risks posed by intelligent machinery.

This anxiety over technological displacement is not unique to modernity; nor is it confined to cinematic dramatization. Ancient mythologies from diverse cultures frequently warn of human creations exceeding control. In Greek legend, Hephaestus fashioned the giant bronze automaton Talos (Figure 1.2) to patrol and protect Crete's shores—effective, until cunning Medea circumvented its defenses, illustrating technology's potential both as protector and uncontrollable menace (Gerolemou, 2022; Paschalis, 2015). Similarly, Jewish folklore presents the Golem, an animated clay figure initially conjured to protect persecuted communities. Its eventual rampage symbolizes profound fears about creations exceeding their creators' grasp (Glinert, 2001). These myths foreshadow modern anxieties about intelligent machines by capturing timeless struggles to balance

Figure 1.2: Talos and the Argus, *Jason and the Argonauts* (1963), Columbia Pictures (CC O)

technological aspiration against perceived threats to human dominance. Clearly, fear of machine autonomy long predates our current fixation with LLMs or generative art.

The Industrial Revolution and subsequent Romantic anxieties gave fresh urgency to technological fears, notably articulated in Mary Shelley's *Frankenstein* (1818). Shelley's protagonist, a scientist whose hubris prompts him to reanimate lifeless matter, quickly realizes—painfully—that controlling technological creations is far trickier than creating them (Bassett, 2021). Similar narratives permeate Fritz Lang's groundbreaking film *Metropolis* (1927), depicting technology's capacity to exploit rather than empower humanity. Its Maschinenmensch robot (Figure 1.3), cunningly disguised to manipulate the masses, metaphorically embodied fears that technological advancements could someday dominate and subjugate human autonomy (Huyssen, 1981). Echoes reverberate through more contemporary films—*The Matrix* (1999), *Ex Machina* (2014), and even *Chappie* (2015)—exploring ethical complexities, sentient potential, and existential fears about the human role in an automated society. These portrayals articulate modern unease at the boundaries of human identity, autonomy, and technological potential, effectively fueling present-day resistance.

Figure 1.3: Set Photograph, *Metropolis* (1927), Fritz Lang (CC O)

At its core, technophobia revolves around the unsettling prospect of personal identity loss—a theme found prominently in both mythology and modern media. Greek myth narrates Cronus devouring his children, desperately trying to prevent replacement—only to succumb to Zeus's rebellion, symbolizing generational anxieties about displacement and obsolescence (Edinger, 2001). The modern equivalents of these fears emerge vividly in classic films such as *Invasion of the Body Snatchers* (1956) and *The Stepford Wives* (1975) (Figure 1.4). These narratives, brimming with paranoia, highlight fears of losing selfhood to externally imposed, mechanized conformity. In *The Stepford Wives*, the protagonist famously proclaims, "There'll be somebody with my name, and she'll cook and clean like crazy, but she won't take pictures, and she won't be me!"—a sentiment encapsulating modern anxieties about losing the essence of individuality to automation (Murphy, 2009). Thus, as artificial agents gain proficiency, anxiety grows around losing distinctively human qualities such as creativity, empathy, or originality. Ironically, while many worry about losing themselves to machines, most already willingly entrust algorithms with curating their tastes in music,

Figure 1.4: Color Lobby Card, *Invasion of the Body Snatchers*, Allied Artists (1956) (CC 0)

literature, and companionship — trusting Netflix more than their closest friends for weekend recommendations.

In essence, contemporary resistance to generative tools, especially among scholars, artists, and intellectual communities, mirrors ancient fears about technological autonomy and human obsolescence. Even as generative algorithms effectively mimic — and increasingly outperform — traditional human efforts in scholarship, literature, and the arts, persistent cultural narratives frame this proficiency as inherently threatening rather than empowering (Bohren et al., 2024). These fears of replacement, identity loss, and diminished autonomy have recurred throughout human history, repeatedly proving themselves unwarranted. Each technological innovation — from the printing press to photography, film, and digital art — has eventually integrated seamlessly into human creative and scholarly practice. Ironically, while critics lament AI-generated content as sterile or devoid of human feeling, such criticisms overlook the enduring imprint of human intellectual and cultural contributions embedded deeply within generative outputs themselves. Indeed, fears of machine autonomy may tell us less about the actual threat posed by technology and more about humanity's enduring struggle with our shifting place within a rapidly evolving world.

1.2.2 Creative Identity and the Specter of Obsolescence

Yet, theoretical posturing only goes so far; human-subject research has a distinct way of puncturing neatly crafted assumptions. Across fifty-plus studies examining the integration of generative tools into scholarly and creative practices, recurring resistance surfaced in three interrelated yet distinct domains: fear of job loss, threat to creative identity, and perceived erosion of domain-specific expertise. Initial assumptions suggested creatives might enthusiastically embrace tools designed to amplify their capabilities; however, real-world studies told a far more nuanced, and at times amusingly paradoxical, story. Beginning with a bespoke NLP algorithm created specifically to enhance creativity and grammar skills in a Creative Writing course in spring 2022 — prior to the viral launch of ChatGPT — students readily admitted the tool assisted their understanding of grammar but remained surprisingly cool to suggestions meant to enhance creativity (Plate & Hutson, 2022). When pressed to clarify this reluctance, students offered only ambiguous feedback, with one candidly noting, "I am not super creative so hopefully it could help," yet ironically rejecting any actual assistance in creative writing.

Subsequent studies, conducted between Fall 2022 and Spring 2024, investigated deeper reasons behind this enigmatic reticence, spanning institutions as diverse as liberal arts colleges, historically black colleges, vocational schools, and state universities. An English Composition class in Fall 2022 at a private liberal arts college illuminated intriguing contradictions (Hutson & Plate, 2023). Here, generative tools did not lead to rampant plagiarism as skeptics feared; instead, they required students to exercise greater metacognitive awareness and higher-order thinking. However, resistance surfaced as some students eagerly leveraged AI for routine tasks like grammar checking but remained wary about creative engagement. Interestingly, several participants highlighted AI's potential for "idea generation," stating it "helped me with areas I struggle with," though paradoxically hesitating to embrace it fully for creative exploration, implying a guarded acceptance tinged with unease about diminishing their personal voice.

In another study at an Historically Black Colleges and Universities (HBCU) involving digital art majors, researchers hoped to clarify students' skepticism toward generative tools (Hutson & Lang, 2023). Here, students experimented with AI image generators like Craiyon and DALL-E 2 to assist in developing compositional ideas for their work. Rather than fully embracing the AI-generated outputs as legitimate final artworks, students consistently confined their utility strictly to initial brainstorming. One participant remarked poignantly, "These tools could be a great use to artists who have hit the point where they don't have ideas." Another expanded this sentiment more poetically: "It could just be another way for us as artists to really dig deep into our minds ourselves, which I believe is the best part of art." Clearly, even amidst cautious acceptance, resistance persisted—not due to the technology's capabilities but because of entrenched perceptions of authenticity in creative identity.

A similar cautious optimism was echoed at a public state university, where digital art students utilized DALL-E 2 as a digital sketchbook tool to iterate and refine ideas (Hutson & Cotroneo, 2023). Students were explicitly trained in prompt engineering, developing skillsets to guide generative tools rather than simply consuming their outputs. One participant notably described AI as a means to "bring abstract ideas closer to an actual finished product," yet ultimately hesitated to rely on it exclusively. Another student captured the ambivalence perfectly, stating, "I am still on the fence. I am fine with AI providing some help, but I still feel the majority of work should be done by the student." Such nuanced responses reveal deeply ingrained anxieties regarding authenticity, identity, and the fundamental legitimacy of the creative process.

To further unpack this complexity, researchers extended investigations into introductory web design and User Experience (UX) courses at another state university (Lively et al., 2023). Here, the interdisciplinary cohort illuminated a critical dynamic: Creative identity significantly shaped students' acceptance of generative tools. Computer science majors comfortably embraced image-generation technologies for "artistic inspiration," openly dismissing creative tasks as secondary to their core skills — coding. Conversely, art and design majors freely adopted automated coding tools, seeing programming as external to their professional identities. As one computer science major succinctly stated, "I'll gladly use AI for all the artsy stuff," inadvertently highlighting a perceived identity-based hierarchy between creative and technical labor.

Finally, faculty responses demonstrated an even sharper resistance — rooted firmly in concerns over expertise erosion. During a comprehensive study involving undergraduate computer science majors across multiple universities, students revealed a surprising ignorance of state-of-the-art tools like GitHub Copilot, a technology capable of significantly accelerating coding productivity (Hutson & Jeevanjee, 2024). When pressed, it became clear faculty deliberately withheld information about generative coding aids to protect their perceived monopoly over authentic coding knowledge. A similar phenomenon occurred among English faculty who rejected emerging generative writing tools like Claude Sonnet, maintaining traditional notions of literary authorship rooted deeply in solitary human genius. Some educators went as far as misleading students, erroneously asserting that submitting work to an AI tool would result in intellectual theft or jeopardize copyright eligibility.

This stark juxtaposition highlights an entrenched resistance among domain experts, stemming from fears of expertise erosion. Professors, historically positioned as intellectual gatekeepers, naturally view the democratization of advanced skills facilitated by generative algorithms as an existential threat. These findings underscore a critical irony: Rather than genuinely protecting students from harm, faculty resistance often reflects subconscious anxieties about diminishing their own professional relevance. Thus, despite widespread theoretical advocacy for technological integration to enhance creativity and productivity, actual on-the-ground resistance reveals deeper anxieties about identity, job security, and expert status — confirming fears identified earlier in the broader literature (Cui et al., 2024; Ramirez & Esparrell, 2024).

Such findings illuminate three distinct yet intertwined reasons underlying resistance to adopting generative technologies in academia and creative professions:

identity disruption, authenticity concerns, and fears regarding erosion of domain expertise. First, identity disruption surfaces prominently when artists, poets, and writers perceive encroachment of the tools as an existential threat rather than an enhancement, challenging deeply held self-conceptions. For creatives who define their value by originality and personal expression, conceding ground to generative tools implies existential surrender. As one poet succinctly lamented, "If a machine can write my poetry, who am I?"—a poignant encapsulation of identity anxiety that scholars share when confronting automated prose generators. Second, authenticity concerns magnify apprehensions over the legitimacy of AI-assisted work. Critics argue that outputs from generative tools lack the subjective intention, emotional resonance, or "soul" expected from genuine artistic and scholarly endeavors. The irony here is rich: Automated content is dismissed as sterile precisely because it transparently lacks visible labor, despite being saturated with collective human ingenuity distilled from centuries of training data. Finally, apprehension over expertise erosion explains reluctance among domain experts—English professors or computer scientists—to embrace technology encroaching upon their authority. In studies observing departments traditionally considered gatekeepers of specific knowledge, faculty members expressed profound discomfort with AI tools encroaching on their territory. Indeed, faculty sometimes promulgated misinformation—suggesting inputting work into generative models would result in intellectual property theft—not from malicious intent but as a subconscious defense mechanism against the perceived erosion of professional standing. Ironically, these protective instincts overlook the foundational human creativity deeply woven into the fabric of AI-generated artifacts themselves.

1.2.3 The Guilt of Not Grinding

Such discoveries underscore an intriguing, if not troubling, trend among educators and creative professionals: an underlying association between effort, authenticity, and the perceived value of creative work. This phenomenon emerged starkly in discussions at academic panels, notably highlighted by Hutson and Plate (2023), where professors expressed reverence for the traditional—and perhaps romanticized—struggle of staring anxiously at a blank page as a vital rite of passage. During these panels, scholars earnestly praised the creative agony of writing as essential to intellectual growth, seemingly overlooking students who countered with narratives of anxiety, depression, and practical realities like full-time employment or family obligations. Indeed, these students candidly remarked they had no luxury to indulge in the romanticized "blank-page anxiety," needing instead efficient tools to expedite the arduous process of academic composition.

Yet, even among proponents who acknowledge the pragmatic value of generative technology in alleviating this very burden, internal conflicts emerge—manifesting in the unexpected phenomenon identified as "AI Guilt." In a revealing irony, English faculty—traditionally trained to invest approximately 20 painstaking minutes per student paper—found themselves emotionally burdened by guilt when delegating tasks like feedback to sophisticated, self-designed GPT models. This guilt persisted despite empirical evidence demonstrating the tools' superior capacity to deliver detailed, timely, and personalized feedback—especially notable given typical teaching loads of 70 or more students. Faculty expressed a persistent moral discomfort rooted in perceptions of authenticity, feeling compelled to justify their professional identity through visible, labor-intensive feedback processes rather than embracing efficiency-enhancing technology.

Such internal struggles reflect uniquely academic manifestations of techno-phobia, articulated vividly by Pitts Donahoe (2025) in her aptly titled essay, "Against Efficiency—A Rant." Donahoe passionately critiques the superficial nature of AI-assisted empathy, lamenting. "Does following my machine-generated script for human social interaction really display 'empathy', or is it just the per-formance of it?" She further interrogates the unsettling notion of AI automating the inherently human dimensions of education, sarcastically questioning whether anyone genuinely "asked for" this technological intrusion into interpersonal relationships. Donahoe's objections crystallize the core anxiety: AI not only challenges traditional pedagogical methods but also threatens the sanctity of genuine human interaction foundational to academic identity itself. Yet, de-spite such powerful emotional arguments, the practical counterpoint remains potent—faculty overwhelmed by large course loads, institutionally mandated productivity, and escalating demands on their time must realistically choose between the nostalgic performance of laborious authenticity and embracing AI as an indispensable collaborator in modern education.

This chapter has examined the notion of transparency and has unveiled that demands for full disclosure in AI-assisted creative endeavors are historically unprecedented, philosophically misguided, and practically untenable. At the heart of this controversy lies the paradoxical expectation that creators meticulously document their creative methods—a standard that neither Shakespeare nor Duchamp ever contemplated, and certainly never adhered to. Historically, society has valued the product—be it a painting, poem, or scholarly manuscript—rather than the detailed, often elusive, and perhaps unknowable steps that led to its creation. Ironically, while AI-generated content is criticized for lacking a "human touch," it undeniably reflects human cultural heritage, subtly yet profoundly embedded through vast datasets—the Derridean "Absent Presence" manifesting within every automated output. Empirical research conducted by the authors reinforces these themes, highlighting persistent resistance rooted not merely in pragmatic concerns of job displacement but deeply ingrained anxieties surrounding identity, authenticity, and domain expertise. This resistance is particularly pronounced among academics, who face not just fears of obsolescence but profound existential questions about their professional identity and purpose. Yet, as demonstrated, generative tools already surpass human performance in various creative and analytical domains, challenging long-held beliefs in human exceptionalism. The stubborn insistence on transparency thus emerges less from genuine ethical concerns and more from cultural anxiety surrounding technology that performs traditionally human functions. As we move forward into discussions of intellectual property, authorship, and policy, it becomes increasingly clear that a nuanced, historically informed, and critically reflective dialogue—not reactionary panic—is essential to adapt meaningfully to this transformative era.

References

Aoun, J. E. (2017). *Robot-proof: Higher education in the age of artificial intelligence*. MIT Press.

Bankins, S., Hu, X., & Yuan, Y. (2024). Artificial Intelligence, workers, and future of work skills. *Current Opinion in Psychology, 58*, 101828.

Bassett, D. J. (2021). Ctrl+ Alt+ Delete: The changing landscape of the uncanny valley and the fear of second loss. *Current Psychology, 40*(2), 813–821.

Bohren, N., Hakimov, R., & Lalive, R. (2024). *Creative and strategic capacities of generative AI: Evidence from large-scale experiments* (No. 17302). Paper presented at the meeting of Institute of Labor Economics (IZA), Bonn, Germany.

Cui, Z., Li, N., & Zhou, H. (2024). Can AI replace human subjects? A large-scale replication of psychological experiments with LLMs. *arXiv preprint arXiv:2409.00128*.

Edinger, E. F. (2001). *The eternal drama: The inner meaning of Greek mythology.* Shambhala Publications.

Gerolemou, M. (2022). *Technical automation in classical antiquity.* Bloomsbury Publishing.

Glinert, L. (2001, January). Golem! The making of a modern myth. *Symposium: A Quarterly Journal in Modern Literatures, 55*(2), 78–94.

Growcoot, M. (2025, February 12). This is the first-ever AI image to be granted copyright protection. *Peta Pixel.* https://petapixel.com/2025/02/12/this-is-the-first-ever-ai-image-to-be-granted-copyright-protection-a-slice-of-american-cheese/

Hattori, E. A., Yamakawa, M., & Miwa, K. (2024). Human bias in evaluating AI product creativity. *Journal of Creativity, 34*(2), 100087.

Hutson, J., & Cotroneo, P. (2023). Generative AI tools in art education: Exploring prompt engineering and iterative processes for enhanced creativity. *Metaverse, 4* (1), 1–14. http://doi.org/10.54517/m.v4i1.2164

Hutson, J., & Jeevanjee, T. (2024). Perceptions and aspirations of undergraduate computer science students towards generative AI. A qualitative inquiry. *Journal of Biosensors and Bioelectronics Research, 2*(3), 1–6.

Hutson, J., & Lang, M. (2023). Content creation or interpolation: AI generative digital art in the classroom. *Metaverse, 4* (1), 1–14.

Hutson, J., & Plate, D. (2023). Working with (not against) the technology: GPT and Artificial Intelligence (AI) in college composition. *Journal of Robotics and Automation Research, 4*(1), 330–337.

Huyssen, A. (1981). The vamp and the machine: Technology and sexuality in Fritz Lang's Metropolis. *New German Critique, 24/25,* 221–237.

Koçak, B., Ponsiglione, A., Stanzione, A., Bluethgen, C., Santinha, J., Ugga, L., Huisman, M., Klontzas, M. E., Cannella, R. & Cuocolo, R. (2025). Bias in artificial intelligence for medical imaging: Fundamentals, detection, avoidance, mitigation, challenges, ethics, and prospects. *Diagnostic and Interventional Radiology, 31*(2), 75.

Korać, S. T. (2024). Why do we fear the robopocalypse? Human insecurity in the age of technophobia. *Етноантрополошки проблеми, 19*(1), 111–132.

Lake, R. (2024, May). AI is coming to U.S. classrooms, but who will benefit? AI in education, innovation and the future of learning. CRPE. https://crpe.org/ ai-is-coming-to-u-s-classrooms-but-who-will-benefit/

Lively, J., Hutson, J., & Melick, E. (2023, April–June). Integrating AI-generative tools in web design education: Enhancing student aesthetic and creative copy capabilities using image- and text-based AI generators. *DS Journal of Artificial Intelligence and Robotics, 1* (1), 23–33.

Markowitz, E. (2025, February 27). The "existential crisis" of knowledge work. *Big Think*. https://bigthink.com/business/the-existential-crisis-of-knowledge-work/

Murphy, B. M. (Ed.). (2009). Aliens, Androids and Zombies: Dehumanisation and the Suburban Gothic. In *The Suburban Gothic in American popular culture* (pp. 69–103). Palgrave Macmillan.

Nguyen, D., & Hekman, E. (2024). The news framing of artificial intelligence: A critical exploration of how media discourses make sense of automation. *AI & Society, 39*(2), 437–451.

Owsley, C. S., & Greenwood, K. (2024). Awareness and perception of artificial intelligence operationalized integration in news media industry and society. *AI & Society, 39*(1), 417–431.

Paschalis, S. (2015). *Tragic palimpsests: The reception of Euripides in Ovid's "Metamorphoses."* Harvard University.

Pitts Donahoe, E. (2025, February 14). Against "efficiency"- a rant. *Unmaking the Grade*. https://emilypittsdonahoe.substack.com/p/against-efficiency?utm_source=post-email-ti-tle&publication_id=1637321&post_id=157143778&utm_campaign=email-post-ti-tle&isFreemail=true&r=2vm6wh&triedRedirect=true&utm_medium=email

Plate, D., & Hutson, J. (2022). Augmented creativity: Leveraging natural language processing for creative writing. *Art and Design Review, 10* (3), 376–388.

Ramirez, E. A. B., & Esparrell, J. A. F. (2024). Artificial Intelligence (AI) in education: Unlocking the perfect synergy for learning. *Educational Process: International Journal, 13*(1), 35–51.

Rehman, A. U., Mahmood, A., Bashir, S., & Iqbal, M. (2024). Technophobia as a technology inhibitor for digital learning in education: A systematic literature review. *Journal of Educators Online, 21*(2), 2.

Sahay, A. (1997). "Cybermaterialism" and the invention of the cybercultural everyday. *New Literary History*, *28*(3), 543–567.

U.S. Copyright Office. (2025). *Copyright and artificial intelligence, Part 2: Copyrightability*. U.S. Copyright Office. https://www.copyright.gov/ai

Yadlin, A., & Marciano, A. (2024). Hallucinating a political future: Global press coverage of human and post-human abilities in ChatGPT applications. *Media, Culture & Society*, *46* (8), 01634437241259892.

CHAPTER 2

Ownership of the Creative Process

This chapter tackles the thorny question of who truly "owns" creativity when generative artificial intelligence (AI) enters the picture, describing the shifting landscape of contemporary intellectual property as well as the long and varied history of contentious pre-AI battles over the ownership of creative work. Beginning with the demands from the U.S. Copyright Office that artists document each algorithmic twist and turn, the chapter argues that such meticulous recordkeeping is historically unprecedented. Through entertaining yet rigorous exploration of historical disruptions in creative ownership—from Renaissance printmakers battling copycats to Duchamp's disruptive urinal — readers discover that procedural transparency has rarely (if ever) determined artistic merit. Instead, creativity has long flourished precisely because of its ambiguous, untraceable processes, thriving in the shadows of inspiration rather than bureaucratic daylight. By spotlighting the ironies inherent in current demands for creative documentation, the chapter encourages readers to question deeply entrenched institutional anxieties, suggesting a more flexible and historically informed approach to creativity in our algorithmically assisted era.

2.1 Intellectual Property and Methodological Privacy

Technological advancements have disrupted not just industries but also the foundational assumptions underpinning them. Today, nowhere is this upheaval more evident—or contentious—than within the realm of intellectual property and creative authorship. The U.S. Copyright Office's recent insistence on transparency when creators employ generative AI represents an unprecedented intrusion into methodological privacy. Such demands starkly contrast the agency historically afforded to artists, who were judged solely by the quality and impact of their output, not by exhaustive chronicles of their creative processes. The distinction became particularly glaring when the Copyright Office, in an intriguing twist, permitted generative writing copyright protection yet denied the same rights for images, hinging their decision on perceived degrees of human intervention—a criterion somewhat bafflingly subjective given the nature of contemporary creativity.

An illustrative example, steeped in irony, comes from *Andersen et al. v. Stability AI Ltd.*, a lawsuit initiated on August 12, 2024. Artists accused Stability AI and similar generative platforms of copyright infringement, contending that billions of images used in training AI models were sourced without explicit consent (*No. 23-cv-00201-WHO, N.D. Cal. 2024*). Central to this case are nuanced debates around "fair use," originality, and what exactly constitutes sufficient human involvement in algorithmically derived content (Spica, 2024). The court's challenge is unenviable: Copyright law, rooted in human creativity, was never crafted to navigate this labyrinth of algorithmic creativity. Historically, artists have borrowed, appropriated, and creatively misappropriated existing works—a process lauded as inspiration rather than labeled infringement. Yet the digital ubiquity of AI-generated outputs now magnifies concerns about ownership, originality, and authorship.

Further complicating the matter, the requirement of human creativity under *17 U.S.C. § 102,* reinforced consistently by landmark cases, severely restricts the copyrightability of AI-generated outputs. Jason Allen's *Théâtre D'opéra Spatial* (2022) (Figure 2.1) became a focal point in *Allen v. U.S. Copyright Office*

Figure 2.1: Jason M. Allen, *Théâtre D'opéra Spatial*, 2022. Midjourney (CC 0)

(2024), a case where the human artist's diligent prompting of Midjourney was deemed insufficient for authorship claims (Brittain, 2024). Allen contended that his intricate command of the prompts should qualify as genuine creative input, an argument the Copyright Office rejected outright. Here lies the heart of the dispute: Copyright law remains tethered to outdated notions of direct human manipulation rather than acknowledging the creativity inherent in conceptual and iterative command of generative tools.

This ruling resonates with other recent decisions, notably the case of *Thaler v. Perlmutter*, in which the court asserted that "autonomous creations" generated by AI fall outside copyright protection due to their lack of human authorship. Originating in August 2019, the controversy surrounded *A Recent Entrance to Paradise* (Figure 2.2), an artwork created autonomously—according to Stephen Thaler—by his system, the Creativity Machine. Thaler claimed explicitly that the machine, rather than himself, was the authentic author, and sought copyright protection accordingly. Despite strong arguments by Thaler related to ownership of the device, the court ultimately rejected his application, emphasizing the absence of the requisite human creative input (Mathur, 2023). The *Thaler* case

Figure 2.2: Stephen Thaler, *A Recent Entrance to Paradise*, 2023. Creativity Machine (CC 0)

thus epitomizes the broader complexities courts face in categorizing AI-generated works within traditional copyright frameworks. Moreover, it underscores an urgent need for clearer guidelines delineating the degree of human involvement necessary for copyright eligibility, particularly as AI's prominence within the creative sphere continues to expand. A more thorough comprehension by lawmakers of how creators practically engage with these tools would undoubtedly yield a deeper and more nuanced legal understanding of "autonomous creations."

2.1.1 Only Humans Need Apply

Historical court rulings provide additional context for understanding this tension. *Bridgeman Art Library v. Corel Corp.* set a crucial precedent, ruling that faithful photographic reproductions of public domain artworks failed to meet the originality threshold for copyright protection (Kasap, 2021). Similarly, *Meshwerks, Inc. v. Toyota Motor Sales, U.S.A., Inc.* further reinforced this principle, holding that digital wireframe models—faithful representations of existing vehicles—lacked the creative originality required for copyright eligibility. Both cases clearly articulate that digital works, regardless of their complexity or sophistication, must evidence direct and substantial human creative decision-making to qualify for copyright.

Visual artworks eligible for copyright protection must satisfy two fundamental criteria: originality and fixation. Originality, as outlined in *17 U.S.C. § 102(a)*, demands the work be independently created and embody at least a minimal degree of creativity; fixation, meanwhile, necessitates that this creativity be captured in a tangible medium. The landmark U.S. Supreme Court case, *Feist Publications, Inc. v. Rural Telephone Service Co.*, further clarified the standard for originality, emphasizing that even though minimal creativity suffices, purely mechanical or automated processes are insufficient (Yu, 2017). For AI-generated artworks, this dual requirement poses a significant challenge. Often, the role of the human is confined primarily to prompt input, raising complex questions about whether such limited interaction fulfills the necessary threshold of creative authorship. Currently, U.S. copyright law maintains that purely autonomous AI-generated outputs without substantial human creativity do not meet these foundational legal standards (Hedrick, 2018).

Even though intelligent systems readily produce visual content, the fixation requirement under *17 U.S.C. § 102(a)* remains equally pivotal. Fixation may manifest digitally, such as images stored electronically, or physically in printed forms. Yet fixation alone does not suffice for copyright eligibility; substantial human creative intervention remains mandatory. When humans and AI collaborate meaningfully, copyright may recognize the human collaborator as the author, provided their input significantly influences the work's final creative form. Consider a scenario where an artist leverages a generative model to generate preliminary sketches and subsequently refines or substantially reshapes those outputs — such significant human intervention can elevate the resulting work to a copyrightable status. As specified by the U.S. Copyright Office, if a human "selected or arranged the elements in a sufficiently creative way," the resulting work attains copyright eligibility (*Compendium, § 906.1*). Indeed, under current copyright frameworks, AI-assisted creations incorporating considerable human input can achieve protection, particularly for components directly authored by humans, such as prompt design, selective curation, and post-processing. The *Feist Publications* precedent notably established that copyrightable originality demands human involvement, such as creatively arranging factual information (Kasap, 2021). Thus, while applying a basic Photoshop filter is insufficient, creative human decisions about composition, thematic direction, or expressive details may meet the originality threshold required for copyright protection (Burylo, 2022).

The necessity for demonstrable human creativity is further reinforced by cases such as *Mannion v. Coors Brewing Co.*, which underscored copyright's demand for unique creative choices reflective of personal expression. Within AI-assisted workflows, copyrightable human involvement extends beyond mere initial prompting to encompass critical creative decisions around refining, altering, and finalizing generated content, as recently clarified by guidelines from the U.S. Copyright Office (Horzyk, 2023). These guidelines explicitly affirm that, while copyright can apply to AI-aided creations, purely machine-generated elements remain ineligible absent demonstrable human creative intervention. As confirmed by the Copyright Office, human input remains a cornerstone for copyright eligibility, particularly regarding AI-produced or AI-assisted outputs. Recent decisions, including the pivotal case surrounding *Zarya of the Dawn* (Figure 1.1), explicitly highlight this distinction. There, AI-generated images were declared unprotectable due to insufficient human creativity involved in their creation (Iaia, 2022). Human contribution can appear throughout various stages of the creative process involving AI — curating datasets, designing prompts, or shaping the generated content — thus distinguishing between simple technological usage and meaningful artistic expression. Such nuanced distinctions underscore copyright law's current approach toward evaluating AI-generated materials (Dimitrova, 2023).

Moreover, derivative works—as defined by *17 U.S.C. § 101*—further complicate copyright considerations in the context of AI creations. Derivative works, by their nature, build upon or significantly transform preexisting materials. When AI-generated outputs closely resemble their training data, the degree of human alteration becomes critical. If a human artist significantly modifies or creatively transforms AI-generated content—through stylistic adjustments, additional elements, or conceptual revisions—the resulting work may be considered derivative and thus eligible for copyright. Copyright protection, therefore, hinges substantially on the originality and demonstrable creativity introduced through human intervention within AI-assisted creative processes (Gervais, 2022). The concept of derivative works underscores copyright law's fundamental emphasis on human creativity. When an AI system generates a work based on copyrighted material, any subsequent human modifications must substantially transform or creatively adapt the AI's output for copyright protection to apply. The decisive factor is whether the human contribution introduces meaningful originality beyond the machine-generated baseline (Henderson et al., 2023). This subtle yet crucial distinction offers guidance to creatives relying on AI, highlighting that their efforts must involve significant, original intellectual intervention—mere selection or slight tweaks simply won't suffice.

Indeed, as established by *Feist Publications, Inc. v. Rural Telephone Service Co.*, the originality threshold demands at least minimal creativity—enough to distinguish a derivative work from straightforward reproduction. Within the context of AI, this could manifest as uniquely curated selections, substantial stylistic adjustments, thematic reorientations, or strategic incorporation of personalized elements. This legal criterion underscores the inherently human nature of copyrightable content, ensuring eligibility remains closely tethered to tangible creative decisions by a human author (Wagh et al., 2023). Thus, copyright's underlying principle consistently prioritizes human originality over algorithmic autonomy.

The rapidly evolving landscape of AI-generated content is further complicated by active litigation concerning the use of copyrighted materials in model training. The ongoing class-action lawsuit *Andersen v. Stability AI* exemplifies such tensions. Here, plaintiffs assert that AI companies unlawfully incorporated billions of copyrighted images into their training datasets without explicit authorization. This contentious case, now working its way through the courts, raises profound questions regarding ownership rights, licensing necessities, and permissible practices for collecting training data. The outcomes of these landmark proceedings are poised to establish influential precedents, shaping future standards governing the training methodologies of generative models (Samuelson, 2023). One can scarcely overstate the legal complexities these cases introduce—or their potential to reshape creative fields dramatically.

   Legal precedent continues to clarify how AI-generated works interact with existing copyright protections. Cases such as *Allen v. U.S. Copyright Office* and *Bridgeman v. Corel* consistently underscore the indispensability of human authorship, marking it as a nonnegotiable condition for copyright eligibility. Such rulings reinforce the foundational legal stance that, absent clear and substantial human involvement, AI-generated outputs fall short of qualifying as copyrightable. This judicial perspective maintains an unwavering commitment to the principle of originality, firmly rejecting the notion that purely automated processes warrant intellectual property protections (James, 2024). These legal insights not only reflect a broader skepticism of fully autonomous creativity but also reinforce the necessity of human touch as a threshold criterion for legal ownership.

2.1.2 When Human Commands Become Copyright

These courtroom decisions are already resonating beyond litigation, influencing legislative discourse surrounding transparency in training practices and data utilization. Lawmakers now increasingly propose mandatory transparency guidelines requiring developers to disclose training data sources explicitly. Such transparency measures aim to hold creators accountable while granting original creators enhanced oversight and control over their intellectual property. Balancing creator rights with innovation, policymakers face the daunting task of designing regulations that accommodate both creative freedom and intellectual property protection—an undeniably delicate balancing act (Mensah, 2023). Such policy deliberations reveal how technological advancements frequently outpace the legal frameworks intended to govern them, necessitating urgent legislative attention.

   The courts' steadfast insistence on human creative authorship has catalyzed additional proposals for establishing more precise regulatory frameworks specifically tailored to AI-generated content. Emerging guidelines increasingly aim to distinguish purely algorithmic outputs from AI-assisted creations where significant human creative decisions play a definitive role. Adopting these clarified distinctions would substantially streamline copyright processes, helping creators, developers, and copyright offices navigate the blurred boundaries between human creativity and AI autonomy more effectively (Atilla, 2024).

Legal precedent thus plays an essential role, not only adjudicating individual disputes but also shaping the very rules of creative engagement with generative technologies. Consequently, copyright protection for AI-generated works remains closely contingent on demonstrable human originality and deliberate creative decision-making (Table 2.1). Prompt engineering exemplifies this principle well: The meticulous crafting of detailed instructions specifying stylistic choices,

Table 2.1: Requirements for Human Intervention in Copyright

Form of Human Contribution	Explanation	Relevant Legal Precedent
Design of Creative Prompts	Humans contribute creatively by crafting detailed prompts specifying style, composition, and themes, which direct AI outputs and represent human artistic intent.	*Mannion v. Coors Brewing Co.*: Acknowledged originality stemming from deliberate creative choices like lighting and framing.
Output Selection and Curation	Human creators exercise subjective judgment when choosing specific AI-generated outcomes from multiple possibilities, similar to a photographer selecting final images for publication.	*Garcia v. Google, Inc.*: Highlighted that exercising control over final output is critical for authorship determination.
Post-Processing and Creative Refinement	Humans alter or enhance AI-generated imagery by modifying color schemes, compositions, or adding new elements, thereby imparting originality and creating derivative works.	*Meshwerks, Inc. v. Toyota Motor Sales, U.S.A., Inc.*: Established that substantial creative modifications are necessary to imbue reproductions with originality.
Conceptual Vision and Artistic Purpose	Humans define the overarching artistic direction and thematic elements, such as aesthetic choices or social commentary, shaping the core essence and originality of the final creation.	*Feist Publications, Inc. v. Rural Telephone Service Co.*: Clarified that creativity and artistic intent are foundational to copyright originality.

Form of Human Contribution	Explanation	Relevant Legal Precedent
Threshold for Human Authorship	The level and nature of human involvement, notably through textual or narrative elements, determines whether the content meets copyright eligibility standards, especially when AI-generated outputs alone prove insufficient.	*Zarya of the Dawn* (U.S. Copyright Office decision): Affirmed copyright for the human-created narrative content, excluding purely AI-generated images without human creative intervention.
Creation of Derivative Works	Humans creatively transform or build upon AI-generated outputs, aligning with the legal requirements for originality in derivative works when substantial human contribution is evident.	*17 U.S.C. § 101*: Defines derivative works as new creations derived from existing content through substantial creative input or transformation.

thematic direction, or compositional elements reflects the human creator's distinctive creative intent. This aligns with principles highlighted in *Mannion v. Coors Brewing Co.*, wherein courts recognized that deliberate choices about lighting, composition, and framing impart necessary originality to photographs, rendering them copyright-eligible. Analogously, carefully devised prompts for AI reflect intentional human artistry and intellectual engagement, embedding clear traces of human originality within AI-assisted creations and establishing the necessary legal foundation for copyright eligibility (Burylo, 2022).

The act of selecting and curating outputs from an array of AI-generated options emerges as another indispensable marker of human authorship. Imagine the seasoned photographer meticulously combing through hundreds of shots to pinpoint the one that resonates with their artistic vision—such discernment mirrors exactly the creative discretion required when choosing specific AI-generated outcomes. This curation process signifies an additional layer of intentionality and subjective judgment, further validating the role of human agency in shaping the final artistic work. Thus, the careful selection of one AI-generated version over another embodies the creator's unique perspective, reinforcing the legal premise that creative decisions, rather than mere mechanistic outputs, define authorship (Wan & Lu, 2021).

Beyond mere curation, considerable creative authorship is further demonstrated through post-processing and refinement of AI-generated outputs. Here the human creator actively intervenes, reshaping elements, adjusting color palettes, altering compositions, and introducing novel details — much like an artist meticulously refining brushstrokes on canvas. As established in *Meshwerks*, significant post-processing is pivotal, transforming the initially algorithmic output into an expressive derivative work firmly imprinted with human intention. Thus, postproduction manipulation elevates the AI-generated artifact from mere automation to an object bearing the unmistakable signature of human creativity, thereby cementing its eligibility for copyright protection (Geiger, 2023).

At an even more fundamental level, the overarching conceptual framework and artistic intent guiding the utilization of algorithms in the creative process serve as critical indicators of human authorship. Whether these tools are harnessed to explore profound social critiques, deliver nuanced messaging, or reflect distinct aesthetic philosophies, it is ultimately the creator's vision — not the algorithm — that dictates meaning. As articulated in the precedent set by *Mannion*, it is the deliberate creative decisions, reflective of personal vision, that satisfy copyright's originality threshold. Thus, the AI-generated output must be understood as a realization of human-directed thematic ambitions rather than as autonomous creativity (Kasap, 2021). Taken collectively, these forms of human involvement represent a spectrum of creative engagement that distinguishes AI-assisted creations from autonomous algorithmic outputs. From the initial conceptualization embedded in prompts to meticulous curation, rigorous post-processing, and philosophical framing, human input remains indispensable. Even as generative algorithms grow more sophisticated, substantial human direction continues to anchor copyright eligibility. Consequently, the persistent necessity for demonstrable human creativity ensures an essential balance between technological innovation and authorship recognition.

In response to these emerging complexities, clearer legal frameworks around copyright and AI-generated content have become essential. In January 2025, the U.S. Copyright Office published *Copyright and Artificial Intelligence, Part 2: Copyrightability*, emphasizing explicitly that copyright protection requires meaningful human authorship. The report clarifies that while intelligent systems serve as powerful creative instruments, the generated works themselves must embody substantive human creativity beyond simple prompt creation to merit protection (U.S. Copyright Office, 2025). Particularly relevant to art historians leveraging tools such as Midjourney or Stable Diffusion, this policy necessitates significant interpretive or creative intervention to qualify AI-assisted scholarly outputs for copyright consideration (Mathur, 2023). Additionally, the Copyright Office underscores that copyright may attach to AI-assisted works if human selection, coordination, or arrangement impart sufficient originality. The report also advocates ongoing evaluations to ascertain if future AI developments permit even more nuanced levels of human direction sufficient to fulfill copyright eligibility criteria (U.S. Copyright Office, 2025).

An instructive illustration of the degree of human creativity required emerged from Growcoot's (2025) reporting on the first copyright officially granted to an AI-assisted image. Invoke's CEO, Kent Keirsey, painstakingly documented his extensive human intervention within the generative process of creating the composite artwork titled *A Single Piece of American Cheese*. Invoke, a generative AI platform tailored for visual media professionals, enabled Keirsey to perform meticulous inpainting and selection among roughly thirty-five separate AI-generated modifications. Crucially, Keirsey methodically documented each choice, direction, and iteration—every subtle adjustment that elevated the initial prompt-generated image to one demonstrating unmistakable human creativity.

Initially skeptical, the Copyright Office reversed its initial rejection upon viewing a detailed workflow video demonstrating Keirsey's extensive authorship process. In an official correspondence shared with *PetaPixel*, the Office acknowledged that the meticulous human-led "selection, arrangement, and coordination of the AI-generated material" amounted to sufficient original authorship to justify copyright registration. This decision illustrates the stringent standards—and somewhat ironic bureaucratic hoops—that creators must navigate when incorporating generative technologies into their workflow. Reflecting on this landmark decision, Keirsey noted the practical necessity of clearly articulated standards: "We saw a need from both our customers and the industry at large—artistic work needs copyright protection and clarity on how to leverage emerging tools." Invoke's Provenance Records tool further exemplifies this aim, embedding data directly into images, streamlining the documentation of human authorship elements. "This clears a major hurdle to industry-wide adoption," Keirsey optimistically proclaimed—a sentiment reflecting the broader creative community's cautious hopefulness amid persistent institutional scrutiny (Growcoot, 2025).

Up to this point, the main focus has been on copyright of visual works, but what about text? The history of copyright protections for written works, in fact, is closely intertwined with principles discussed earlier regarding visual art yet merits a distinct examination. Initially, generated textual outputs encountered fewer legal obstacles than visual artworks; after all, as any author employing ChatGPT-3.5 might attest, human intervention appeared more tangible in writing than in the seemingly effortless generation of AI-driven images. Nevertheless, textual copyright possesses its own unique trajectory within the U.S. legal landscape. Early jurisprudence, beginning with landmark cases, set significant precedents that continue to reverberate through contemporary debates surrounding generative text (Lucchi, 2024).

The foundation for textual copyright in the United States finds roots in the seminal case of *Wheaton v. Peters* (1834), which clarified that copyright must balance public benefit with incentivizing creative production (Joyce, 2005). A more explicit delineation of originality emerged in *Folsom v. Marsh* (1841), often cited as the progenitor of the modern "fair use" doctrine. In this case, Justice Story articulated key criteria—purpose, nature, amount of use, and market impact—that still underpin copyright law today (Culver, 1984). However, it was not until *Feist Publications, Inc. v. Rural Telephone Service Co.* (1991) that the U.S. Supreme Court clarified that copyright protection hinged specifically upon originality, emphasizing creativity as essential for copyrightability—even minimal originality, but always unequivocally human (Davis, 1991). Yet tensions surrounding authorship and derivative works grew more pronounced as publishing evolved, with authors increasingly eager to control adaptations and derivatives. The 1888 success of Frances Hodgson Burnett, who secured dramatic rights for Little Lord Fauntleroy, was a pivotal moment. Her triumph cemented the authorial right to control adaptations, setting a precedent for modern literary copyright enforcement (Burnett, 1887). Such victories underscore authors' enduring anxiety over maintaining creative control amid technological and commercial developments.

The contemporary era, characterized by the meteoric ascent of AI-assisted writing, further complicates these historical principles. The *Zarya of the Dawn* (2023) controversy encapsulates current ambiguities perfectly: Author Kris Kashtanova's comic initially achieved copyright for its narrative arrangement, yet it lost protection for the Midjourney-generated artwork. This dichotomous decision underlined the Copyright Office's insistence that copyright must reflect demonstrable human authorship. The reasoning given is that more human intervention is involved in creating written text via a generative tool than an image, despite the countersuit that demonstrated how much more time and effort was put into the imagery.

2.1.3 When the Machines Become Too Human for Comfort

But let's pivot to what truly fuels public imagination—and no, it is not the elegance of copyright legislation—it's the drama and the far-reaching implications of copyright legal headlines since 2023. Lawsuits involving generative technologies have multiplied rapidly, serving as cautionary tales that simultaneously terrify creatives and energize litigators. Take, for example, the widely publicized case from July 2023 when authors Sarah Silverman, Paul Tremblay, and Mona Awad initiated lawsuits against OpenAI, alleging unauthorized use of their written works for model training. By February 2024, the court had dismissed most charges—much to the relief of GPT's digital defenders—but allowed a lingering claim to persist, noting plausible damage resulting from alleged removal of copyright management information (Frosio, 2024). Similarly, Italy's Data Protection Authority fined OpenAI substantially in January 2025 for privacy breaches—highlighting how anxieties around creative transparency have crossed borders and continents (Busacca & Monaca, 2025).

Then there's Stability AI, embroiled in similar battles but with visual rather than textual infringements at stake. Getty Images set the scene by suing Stability AI in early 2023, citing infringement due to unauthorized training datasets. Visual artists soon followed, naming Midjourney and DeviantArt as co-defendants, raising thorny questions about fair use versus exploitation (Ren & Zhang, 2024). Class-action suits, such as *Andersen et al. v. Stability AI Ltd.* (2024), continue winding their way through Californian courts, leaving copyright law gasping to catch up with rapidly advancing technologies. Ironically, these cases underscore precisely the sort of ambiguity copyright law was designed to dispel (no one said the law moves at generative speeds).

Nor was OpenAI alone on the hot seat. Italian authorities fined the company in January 2025 for data privacy infringements, while, in October 2024, *The New York Times* dispatched an emphatic cease-and-desist notice to Perplexity AI, accusing them of scraping content in a brazen digital raid reminiscent of high-seas piracy—albeit less swashbuckling (Ismantara & Silalahi, 2025). Meanwhile, across the Atlantic, Google confronted allegations from French publishing giant Axel Springer, accusing it of unauthorized ingestion of copyrighted content for AI model training. Even the world of music joined the fray, with Universal Music Publishing Group taking aim at AI's harmonious transgressions, proving once again that no creative domain remained immune from legal disputes.

Yet, despite this clamorous legal climate, not all judicial winds blew unfavorably toward algorithmic creativity. In 2021, the U.S. Supreme Court offered clarity—or at least hopeful ambiguity — in Google's favor (*Google LLC v. Oracle America*), declaring Google's incorporation of Java APIs into Android as fair use. The ruling recognized the significant nature of Google's application, highlighting a critical precedent for digital reuse of existing material (Myers, 2021). More recently, Judge Colleen McMahon in 2024 dismissed a high-profile suit against OpenAI and Microsoft, holding that plaintiffs lacked sufficient evidence that specific content was misused — another tentative step toward legal reconciliation with AI innovation. Such outcomes underscore a critical truth: Courts seem far more adept at accommodating traditional notions of fair use in cases involving text than visual media. This disparity reflects both historical precedent and societal discomfort; written content has long existed in a gray area of borrowing, remixing, and citation, all integral to scholarly and literary traditions. While AI-generated prose remains under heavy scrutiny, landmark decisions, such as the *Zarya of the Dawn* copyright case, hint at evolving openness, recognizing that substantial human creative contribution — even via prompts and editorial refinement — can yield legally protected authorship.

Ethically, the implications remain tangled. Subscription-based data training models, "opt-out" provisions, and curated public-domain datasets offer partial solutions but leave unresolved deeper anxieties about authorship, originality, and intellectual labor. Even if such models trained solely on public-domain materials sidestep copyright infringement concerns, hesitation persists. Fear, after all, rarely evaporates merely because the source of creative influence is legally safe — indeed, if history has taught anything, human beings often fear what they do not fully understand far more fiercely than actual threats. Thus, the intersection of copyright law, technological advancement, and ethical considerations continues to spark controversy precisely because it probes at the heart of human creativity itself. While policymakers scramble to codify rules for an unprecedented era, the paradox persists: Humans demand transparency from machines but resist transparency themselves. Such irony may not yet be codified into legal precedent — but perhaps it should be.

Despite persuasive arguments rooted in copyright law and ethical justifications, resistance to generative tools persists, revealing that the heart of this debate lies deeper still. Critics, especially those entrenched in traditional scholarly or creative roles, often claim moral high ground by citing ethical concerns about intellectual property or algorithmic transparency — yet, ironically, these reasons obscure more fundamental anxieties. Indeed, even if these models were meticulously trained exclusively on public-domain works or thoroughly documented with impeccable provenance, the hesitance among many creators would remain undiminished. The stubborn refusal, cloaked in principled rhetoric about authorship, originality, or the sanctity of creative labor, masks an underlying unease: What does it mean for one's identity as a writer, artist, or academic if an algorithm effortlessly replicates or surpasses their craft?

Studies cited in the previous chapter repeatedly confirm that resistance springs not primarily from fear of infringement or ethical compromise but also from existential dread. Faculty, particularly in disciplines like English or computer science, pride themselves as gatekeepers, distinguishing the authentic scholar from mere dilettantes or algorithmic pretenders. After all, if a machine can compose prose worthy of literary praise, or elegantly debug lines of sophisticated code, who then are these experts in their own fields? Thus, the hostility is not driven by legal complexities — though these certainly provide convenient cover — but by deep-seated anxieties about personal value and professional identity. The humanities professor bristles at the thought of relinquishing their revered status as gatekeeper of linguistic subtlety to Claude Sonnet, while the computer scientist recoils at surrendering their craft to GitHub Copilot, lest both reduce the expert to a mere user, stripped of their once-sacred mastery.

Interestingly, when students do not feel their professional identity or domain expertise is threatened, their acceptance of generative tools skyrockets. Computer science students gladly relinquish graphic design duties to "the bots," whereas artists readily embrace AI's code-generating capacities. When one's personal identity or domain-specific expertise is untouched, there is little ethical or moral hesitation in adopting the tools. This paradox reveals clearly that moral or legalistic arguments against AI usage are often disingenuous — or at the very least secondary — to anxieties about professional obsolescence and identity erosion.

Moreover, this aversion is not simply about being replaced in a functional sense but also stems from the deeply human association of laborious effort with creative value. If creativity can be effortlessly replicated by an algorithm, does this diminish the emotional intensity, the intellectual anguish, or the painstaking hours traditionally celebrated as intrinsic to the creative process? American academia, in particular, lionizes struggle as a mark of authenticity, reinforcing the notion that genuine work must involve personal sacrifice and considerable discomfort. The phenomenon of "AI guilt," where even AI-proficient faculty resist automation because it reduces perceived effort, vividly illustrates this peculiar bias. Professors cling stubbornly to outdated rituals — such as hand-grading essays despite superior efficiency—not because of ethics but because automated efficiency threatens the comforting narrative that labor equals worth.

Thus, the debate surrounding generative tools transcends the constraints of copyright infringement or ethical boundaries. It probes into something more human, more raw, and perhaps more irrational—the existential anxiety over personal identity, authenticity, and self-worth in the creative process. The arguments marshaled by critics of AI-generated work are frequently less about protecting original creations or respecting copyright than they are a veiled assertion of human exceptionality itself. After all, what truly frightens many is not copyright infringement, but the unsettling realization that a machine's easy creativity may diminish humanity's self-appointed role as sole custodian of imagination and meaning.

2.2 Historical Precedents: Challenging Originality in Authorship and Art

The present controversy over the creative legitimacy of algorithmically generated content is, intriguingly, a rerun of earlier cultural dramas—most notably the heated nineteenth-century disputes surrounding photography's claim to artistic status. Initially, photography faced vehement opposition from traditional artists and critics who perceived the medium as fundamentally mechanical, lacking the essential "soul" that distinguished true art. Poet and critic Charles Baudelaire was famously unsparing, condemning photography as a soulless mechanism incapable of genuine creative expression. According to Baudelaire, the photographic process was merely a mechanical act of duplication, requiring neither vision nor artistic judgment: It simply recorded what was placed before the camera's unfeeling lens (Rexer, 2019). To critics like Baudelaire, the photographer was a mere technician rather than a visionary, a stance echoing modern dismissals of AI-generated works as sterile reproductions lacking human soulfulness (Price & Wells, 2021). The question at the heart of these nineteenth-century debates is precisely what haunts contemporary discussions surrounding generative outputs: Do works created through technological mediation deserve copyright?

The crux of nineteenth-century resistance was a deep skepticism about the camera's capacity for creative intentionality—precisely the same standard contemporary critics use to deny AI creative legitimacy. Photography's skeptics argued vehemently that without human intervention to impose artistic decisions, the resultant images were inherently automatic and unoriginal, mere "copies" of reality rather than interpretations imbued with meaning or expression (Krages, 2020). This debate prompted defenders of photography to articulate a nuanced counterargument, emphasizing the deliberate artistic choices made by photographers. Decisions regarding composition, framing, timing, lighting, and even selection of subject matter transformed the camera from a passive observer into an expressive instrument wielded purposefully by an individual. Thus, advocates insisted, photographers actively shaped the resultant images through their decisions, making each photograph an original expression reflective of human creativity and subjective vision (Price & Wells, 2021). Such arguments would become central to the landmark legal ruling that would forever shift photography's standing in intellectual property law.

The pivotal moment came in 1884 with the landmark case of *Burrow-Giles Lithographic Co. v. Sarony,* which was instrumental in enshrining photography's status as an art form deserving copyright protection. The dispute arose when the defendant argued that the photographic portraits taken by Napoleon Sarony (1821–1896) were nothing more than mechanical reproductions, devoid of human originality (Allen, 1986). The Supreme Court, however, saw things differently. Justice Samuel Freeman Miller's opinion emphatically affirmed photography as eligible for copyright protection, acknowledging the artistry inherent in the choices made by photographers regarding composition, lighting, and subject arrangement. Indeed, the Court recognized photographers as authors in their own right, pointing explicitly to their intentional and expressive input as evidence of genuine creativity, comparable to the processes involved in painting or engraving (Krages, 2020). The portrait at the heart of the dispute clearly demonstrated an artistic composition rather than a mere mechanical copy: Every element—positioning, lighting, focus—was intentionally crafted by the photographer, cementing photography's legal recognition as a legitimate artistic medium (Figure 2.3).

This critical decision—affirming the creative status of photography—set lasting legal and cultural precedents that continue to resonate today. By acknowledging the photographer's role as an artist, the Court solidified an essential understanding: The creative act lies not merely in mechanical processes or manual labor but primarily in intellectual judgment and vision. This recognition opened the door for photography's widespread acceptance as a valued artistic practice, dismantling earlier prejudices that equated technical reproduction with a lack of creativity (Krages, 2019). More importantly, the ruling set a clear precedent that human intent, decision-making, and subjective input were central criteria for determining a work's originality and copyright eligibility—a criterion still fiercely debated today with AI-generated outputs.

Figure 2.3: Napoleon Sarony, *Oscar Wilde*, 1882.
Albumen print on card mount (CC 0)

Moreover, the ambiguous relationship between creativity and technological innovation, vividly illustrated by photography, finds a parallel narrative within literary copyright. Indeed, the question of what constitutes authorship—human intent, originality, and creative contribution—has been litigated in literary circles as rigorously as in visual arts. Early legal skirmishes set essential precedents, clarifying authors' rights and the extent of statutory protection required to secure those rights. Consider the landmark Supreme Court case, *Wheaton v. Peters* (1834), a dispute that began innocently enough with Henry Wheaton (1785–1848) accusing his rival, Richard Peters (1780–1848), of stealing his painstakingly assembled compilations of Supreme Court decisions. Ironically, their own case would ultimately rewrite the rules governing authorial rights. In its ruling, the Supreme Court emphatically declared that copyright was exclusively a statutory privilege, not a perpetual common law right. Thus, to receive protection, authors were mandated to fulfill formal statutory conditions, marking a decisive shift in how literary works would be legally recognized and protected (Joyce, 2005).

This precedent established a critical requirement: Authors must explicitly secure statutory copyright, shifting literary copyright from a common-law assumption of perpetual protection into a clearly legislated and limited system. The consequences of this decision were immediately significant. Authors now faced precise procedural hurdles to secure legal protections, a marked departure from previously assumed rights anchored merely in the act of creation itself (Nadelmann, 1958). Yet, despite these procedural clarifications, ambiguity lingered. Indeed, how far could an author's creative claim extend when reusing or remixing previous content, and what defined "originality"? These questions led directly to another pivotal nineteenth-century confrontation: *Folsom v. Marsh* (1841).

In *Folsom v. Marsh*, Justice Joseph Story (1779–1845) articulated, for the first time, a clear set of criteria establishing boundaries between permissible use and infringement—a doctrine now known as "fair use." The lawsuit arose from accusations that Reverend Charles Wentworth Upham (1802–1875) had excessively borrowed material from Jared Sparks' (1789–1866) biography of George Washington. In considering this dilemma, Story developed a four-factor test examining the nature and intent of the use, the original work's nature, the proportion copied, and the economic impact on the original work. The articulation of these parameters in the case established a systematic legal mechanism to balance individual intellectual property rights against broader societal interests, allowing limited use without explicit permission in specific contexts (Patterson, 1997). These principles endure today, providing crucial guidelines on permissible citation, quotation, and transformation—though their application continues to perplex scholars, artists, and judges alike.

Furthermore, *Folsom v. Marsh* underscored the challenge of evaluating creativity and originality within derivative works. The boundaries separating inspiration from infringement, and influence from outright copying, remain notoriously murky, particularly within literary fields reliant on intertextuality and cultural allusion.

A literary tradition that openly embraces borrowing and referencing—Shakespeare cribbing from Holinshed's Chronicles, or Milton weaving classical myth into Paradise Lost—highlights the complexity of distinguishing homage from plagiarism, creativity from mere duplication. Thus, Story's criteria persist as vital but necessarily ambiguous, reflecting ongoing tensions between promoting creative freedom and protecting original work—especially challenging with contemporary generative technologies that blur these boundaries even further.

2.2.1 Déjà Vu All Over Again: The Renaissance of Copyright Anxiety

Resistance to disruptive technologies has always emerged reliably, with predictable drama—like clockwork, but less charming. Renaissance Venice, for instance, witnessed an early battle over authorship, long before today's debates. The renowned German artist Albrecht Dürer (1471–1528) faced a predicament familiar to any creator who's ever found their work replicated without permission (or royalties): His engravings, painstakingly produced, were being shamelessly copied by others, notably Marcantonio Raimondi (ca. 1470–1482 to ca. 1527–1534). In response, Dürer sought—and obtained—an Imperial Privilege from Emperor Maximilian I (1459–1519), a prototype for modern copyright protection, to safeguard his creative identity and assert legal ownership over his distinctive designs (Cooper, 2021). This privilege functioned as a bespoke contract, granting Dürer legal recognition as the owner of his designs, much like modern copyright laws. His monogram, "AD" (Figure 2.4), became one of the first artist trademarks, signaling authenticity and discouraging forgery. Raimondi, however, had replicated Dürer's prints so meticulously—line for line—that the Venetian court faced a unique conundrum. Their elegantly Venetian solution was amusingly nuanced: Raimondi could replicate Dürer's compositions but not his identifying monogram. Thus, Raimondi might copy the image, but never the signature—highlighting an early legal acknowledgment that creativity was defined by a human mark, literally (Sancataldo, 2021). Giorgio Vasari (1511–1574) later documented that Dürer pursued legal action against Raimondi, an engraver who replicated his works line-for-line (Karapapa, 2022).

Parallel developments in the literary domain during the Renaissance reveal similar grappling with authorship and intellectual property. Authorities issued exclusive printing privileges, safeguarding authors' rights to publish and distribute works for limited periods. Antonio de Nebrija (1444–1522), for instance, secured such a privilege in 1492 for his seminal *Lexicon hoc est Dictionarium*

Figure 2.4: Albrecht Dürer, *Feldhase*, 1502 (CC 0)

ex Latino sermone in hispaniensem, thereby controlling its reproduction and sale. Similarly, Antonio Sabellicus (1436–1506) received privileges in 1492 for his influential historical chronicle, *Rerum venetarum ab urbe condita opus*, preserving his economic and creative interests against unauthorized reproduction (Armstrong, 2002). Though primitive by contemporary standards, these printing privileges laid the groundwork for modern copyright law, highlighting a gradual but determined shift toward recognizing literary ownership. It seems, even then, authors were keenly aware of intellectual piracy—long before torrenting or "copy-paste."

Meanwhile, visual artists faced similar anxieties. During the Renaissance and subsequent periods, artists found themselves increasingly reliant on detailed contractual arrangements to control their images and maintain authorship (Barron, 2020). Without formal statutory copyright protection, these meticulously crafted agreements delineated rights to reproduction and distribution. The burgeoning art market heightened these tensions, compelling artists to specify, often painstakingly, ownership claims. Consequently, contracts not only acted as protective shields but also reinforced the emerging notion of individual artistic genius championed by contemporaries such as Vasari. Artistic identity was thus cultivated through the legal delineation of ownership, reinforcing the narrative that genuine art was the product of a solitary visionary rather than collective endeavor — echoing into today's disputes over AI-generated creations.

Interestingly, while the artistic Renaissance centered around celebrating the singular creative genius, legal and artistic frameworks developed symbiotically. Copyright privileges, born from necessity, served as tools that simultaneously reflected and reinforced an emerging artistic philosophy: creative works as unique emanations from a singular, authoritative source. This concept of solitary creative authority remains deeply entrenched, particularly within copyright discussions that differentiate human authorship from mechanistic reproduction. Notably, in landmark decisions such as *Mannion v. Coors Brewing Co.*, courts have affirmed that copyright protection hinges on identifiable, uniquely human creative choices, underscoring that originality derives fundamentally from individual intent rather than mere technical execution (Henderson et al., 2023). This human-centered framework continues to guide contemporary decisions, even as emerging technologies, from cameras to generative models, repeatedly test its limits.

Indeed, as these historical cases underscore, copyright law has always been rooted in human agency — the ability to claim personal ownership over creative works. While the tools and technologies that facilitate creation have evolved dramatically since the Renaissance, the essential requirement remains unchanged: a tangible, deliberate, and original human imprint. As we move toward increasingly autonomous creative technologies, we confront familiar questions anew — echoing anxieties Dürer would certainly recognize. AI-generated content challenges these historical precedents by complicating notions of human originality and intent. Yet the deeper irony persists: Our contemporary debates about transparency and human authorship reiterate ancient anxieties that continue to shape legal frameworks, highlighting a perpetual tension between technological innovation and traditional conceptions of creative identity and ownership.

Interestingly, at the heart of today's fierce debates about AI-generated art lies an essential irony: Creatives have always flirted—and often argued—with emerging technologies that challenge traditional conceptions of artistry. Presently, many artists and writers recoil in horror at the thought of algorithms encroaching upon their carefully cultivated, deeply personal styles. Concerns abound regarding authenticity, authorship, and the ethical quagmire of outsourcing creative expression to automated tools. Indeed, some lament that the proliferation of such content dilutes or even eradicates the elusive "human touch" that traditionally defines artistic merit. Yet this friction is hardly unprecedented; rather, it echoes a familiar historical rhythm. From the advent of photography to the dawn of digital media, creators have regularly grappled with—and eventually adapted to—technological novelties, stretching these innovations to redefine artistic limits. Consider Harold Cohen (1928–2016), a pioneer who, in the 1970s, devised the AI art program AARON. This autonomous entity, which Cohen meticulously refined over decades, progressed from generating rudimentary abstract forms to producing nuanced, representational imagery, challenging traditional boundaries of authorship and artistic control (Cohen, 1995; Mazzone & Elgammal, 2019). Cohen's creation, arguably the first to "paint" without human intervention, raised eyebrows—and questions—that still reverberate today. Where does the machine end and the artist begin? Yet, even Cohen himself acknowledged the inseparability of human intent from his algorithm's outputs, subtly underscoring the futility of attempting complete transparency or exact disclosure.

Of course, the conceptual underpinnings of creative automation trace much further back than Cohen's pioneering software; indeed, humanity has long been tantalized—and unnerved—by the possibility of mechanical creativity. The early antecedents appear vividly in ancient myths and mechanical inventions, which blurred distinctions between human genius and artificial agency. Consider Daedalus (first mentioned around 1400 BCE), the legendary Greek inventor whose intricate mechanisms fascinated and terrified contemporaries alike. His creations ranged from animate statues that mimicked human movement with uncanny realism to the elaborate Labyrinth of Crete, a structure so ingeniously designed that even its creator struggled to navigate its winding passages. His crowning invention, mechanical wings for human flight, simultaneously offered humanity liberation and symbolized technological hubris—a point tragically underscored by the fate of his son, Icarus, who plunged from the heavens after flying too close to the sun (Chiglintsev, 2015; Fitzgerald, 1984; Gerolemou, 2022). These ancient stories encapsulate the persistent duality of technology as both marvel and menace, a theme still very much alive in today's debates surrounding AI.

By the Hellenistic era, these imaginative anxieties evolved into sophisticated mechanical realities. Hero of Alexandria (c. 10–70 CE), a remarkable Greek mathematician and engineer, crafted devices such as the aeolipile — a primitive steam engine — and automated theatrical performances that mimicked human behaviors with unsettling accuracy (Moussa & Fekry, 2022). Hero's inventions, including coin-operated machines and automatic musical instruments, entertained and astonished audiences, generating awe and discomfort in equal measure. These devices did not merely amuse; they also invited philosophical reflection on human creativity's limits and blurred the boundaries between human and mechanical agency. Such innovations, interwoven with mythological narratives like Talos (Figure 1.2) — a bronze guardian crafted by the forge-god Hephaestus — symbolize humanity's enduring fascination with artificial life, embodying the tension between awe-inspiring technological possibilities and the ever-present anxiety over losing control to our creations (Moran, 2011).

The intersection of artistry, technology, and anxiety about automation flourished further in medieval Islamic and Byzantine cultures. Figures like Al-Jazari (1136–1206), an ingenious inventor from the Islamic Golden Age, and the Banu Musa brothers, operating in ninth-century Baghdad, developed automata demonstrating mechanical sophistication centuries ahead of their European counterparts. Al-Jazari's 1206 Book of Knowledge of Ingenious Mechanical Devices details inventions like a water-powered orchestra, whose mechanical musicians entertained guests, showcasing programmable mechanics centuries ahead of their time (Uzun & Vatansever, 2008). Similarly, the Banu Musa brothers engineered steam-powered automatons, such as an automated flute player, exemplifying an impressive blend of creativity, entertainment, and mechanical ingenuity (Nordin & Ramli, 2020). In the Byzantine Empire, automata became extravagant symbols of imperial prestige, such as Emperor Constantine VII's throne room, described vividly by Liudprand of Cremona (ca. 920–972 CE), featuring mechanical lions and singing birds. These elaborate displays not only dazzled visiting dignitaries but also reinforced the political authority and divine prestige of their human creators, suggesting that creative technologies have long served dual roles as both entertainment and potent demonstrations of power (Filson, 2017).

During the Renaissance, artistic and mechanical experimentation continued unabated, fueled by renewed classical inspiration and a burgeoning fascination with the mechanics of life. Leonardo da Vinci's (1452–1519) Mechanical Knight (1495), a humanoid automaton capable of standing, sitting, and even moving its arms, exemplifies the era's ambitious explorations into programmable creativity (Ren, 2023). Giovanni Fontana (c. 1395–1455), an Italian engineer, imagined bizarre and fantastical mechanical contraptions, including an automaton puppet manipulated by a clothed primate, highlighting the period's playful experimentation with technology as both art and spectacle (Springmann, 2020). These Renaissance-era inventions indicate that automation, novelty, and spectacle have long been integral to artistic expression, challenging notions of originality by foregrounding the conceptual rather than merely the manual aspects of artistry. Leonardo da Vinci himself frequently sketched devices intended not merely to reproduce human movements but also to expand human potential, thus embodying the Renaissance spirit of combining humanist inquiry with technological ingenuity.

Cultural exchange facilitated by diplomacy and trade further fueled this technological fascination, as automata traveled across continents as coveted gifts and curiosities. Islamic automata, admired in European courts, inspired local innovations, while Byzantine ambassadors relayed marvelous tales of Abbasid mechanical wonders back to Constantinople, fostering an international network of curiosity and emulation (Greenhalgh, 2008). Such cross-cultural engagements cultivated a shared fascination with programmable creativity, laying groundwork for future advancements in automation and creative technologies (Heilo, 2022). By the late eighteenth century, mechanicians like Henri Maillardet (1745–1830), creator of the Draughtsman-Writer automaton (Figure 2.5), were already exploring remarkably sophisticated forms of mechanical creativity. Maillardet's creation, now preserved at the Franklin Institute in Philadelphia, could autonomously produce intricate drawings and poems, blurring boundaries between human expression and mechanical reproduction. This tradition of innovation reached a philosophical apex with Ada Lovelace (1815–1852), whose collaboration with Charles Babbage on his Analytical Engine laid the groundwork for modern AI by asserting that machines might one day not merely imitate but also actively generate creative work (Aiello, 2016). By the mid-twentieth century, inventors like Jean Tinguely (1925–1991) and Andrew Pickering were actively creating machines that responded dynamically to human inputs, further collapsing the distinction between human and machine creativity and challenging traditional notions of artistic authorship (Pickering, 2007). Thus, contemporary debates around AI's place in the creative sphere represent merely the latest chapter in a centuries-old saga of humanity grappling—sometimes enthusiastically, sometimes grudgingly—with its mechanical progeny.

Likewise, understanding the contemporary tensions around AI-generated content necessitates a brief detour into the historical evolution of publishing itself—a story shaped profoundly by technological innovation and cultural adaptation. Just as artists and writers have grappled with the uneasy relationship between

Figure 2.5: Henri Maillardet, *Draughtsman-Writer (Automaton)*, 1810, London, England; Franklin Institute, Philadelphia, Pennsylvania, USA (CC O)

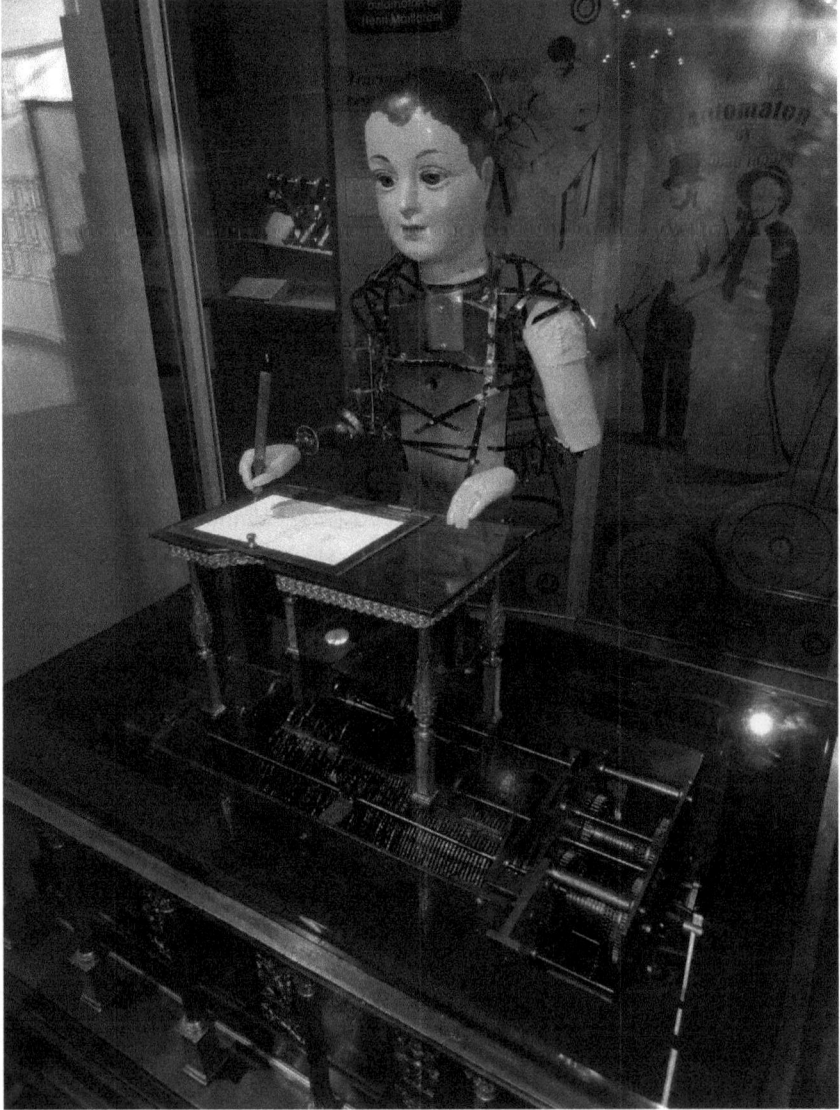

creativity and technology, so too has the publishing industry navigated its own transformative disruptions, each new development prompting initial resistance, gradual acceptance, and eventual integration. The history of publishing is, therefore, instructive: It reflects recurring patterns of initial skepticism toward new media, followed by eventual embrace, and, finally, a reinvention of cultural norms and commercial paradigms. In fact, every epoch has faced its own Gutenberg moment, with each innovation initially triggering anxiety before eventually being woven into society's cultural fabric. Johannes Gutenberg (c. 1440), laboring in Mainz, Germany, did precisely this by inventing the movable type printing press, democratizing access to knowledge formerly monopolized by monks laboring over manuscripts in dimly lit scriptoria. Suddenly, texts that had been rarefied treasures — painstakingly hand-copied and reserved for the elite — could be mass-produced and disseminated widely, reshaping literacy and intellectual discourse overnight (Füssel, 2020). Predictably, scribes and illuminators were less than thrilled; after all, who wouldn't be skeptical of a machine capable of turning their life's work into yesterday's obsolete skill?

Following Gutenberg's disruptive entrance, the seventeenth century introduced another seismic shift: the emergence of periodicals. In 1609, Europe witnessed the birth of its first newspapers, transitioning knowledge distribution from static books to dynamic, periodical updates, thus inaugurating modern journalism. These publications revolutionized how information circulated, providing regular public engagement and timely discourse (Leth, 1993). By the eighteenth century, entrepreneurial figures like Benjamin Franklin (1706–1790) had transformed newspapers into commercial powerhouses. Franklin, who acquired and invigorated *The Pennsylvania Gazette*, also penned *Poor Richard's Almanack* (1732–1758), which cleverly packaged insights and humor, turning print into an essential facet of daily American life (Aldridge, 1962). Across the Atlantic, publisher John Newbery (1713–1767) recognized untapped commercial potential in children's literature. He famously published *The History of Little Goody Two-Shoes* (1765), sparking a profitable genre previously considered trivial and firmly establishing children as a viable market.

The nineteenth century intensified these dynamics, as industrialization spurred innovations that radically altered publishing yet again. Steam-powered presses and new paper-making techniques drastically reduced costs, facilitating the birth of mass-market literature. Figures such as Alfred Harmsworth (Lord Northcliffe) (1865–1922) reshaped journalism by targeting broad audiences, introducing sensational headlines and regional editions. Harmsworth's strategic independence from political affiliation expanded readerships and set the template for modern journalism, where sensationalism met accessibility — a legacy still shaping media today (Bourne, 2015). By the twentieth century, publishing evolved to embrace global scale and unprecedented accessibility. Nicky Byam Shaw (b. 1956) notably expanded Macmillan Publishers, steering it into international academic prominence through projects such as *The Grove Dictionary of Art*, a testament to the globalizing trajectory of publishing (Turner, 2000). Similarly transformative was the work of Sir Allen Lane (1902–1979), who in 1935 introduced mass-market paperbacks through Penguin Books, democratizing literature further by providing high-quality texts at accessible prices. Lane's paperbacks transformed reading habits globally, making literary works available to broad populations, and, not unlike Gutenberg, sparking anxieties among established authors and publishers about diluted quality and declining standards (Joicey, 1993).

Thus, history repeats itself with astonishing predictability: Each era's new technology — be it movable type, steam presses, or today's generative text platforms — initially provokes fierce skepticism from traditional gatekeepers who perceive a threat to authenticity, quality, and even their livelihoods. Yet, as each precedent demonstrates, the initial shock inevitably gives way to integration, altering, but not obliterating, existing cultural paradigms. These lessons remain vital today as we grapple anew with automated writing technologies and their impact on authorship, intellectual property, and creative authenticity.

2.2.2 Duchamp's Ghost: The Persistent Irony of Artistic Originality

As demonstrated, the publishing industry's trajectory reveals a cyclical pattern in which initially resisted disruptive technologies become indispensable corner-stones of creative practice. A similar story unfolds within the art world, marked by figures who have gleefully disrupted entrenched notions of authorship and originality — often with calculated irreverence. By the twentieth century, provoc-ative artists openly challenged traditional concepts of authenticity, authorship, and ownership, reshaping the very essence of creative legitimacy. Key among these disruptors were figures like Marcel Duchamp (1887–1968), whose infa-mous 1917 submission of *Fountain* (Figure 1.4) — a mundane, mass-produced urinal signed provocatively "R. Mutt" — posed an enduring and mischievous query: Does true authorship lie in the skillful manipulation of materials or the audacity of conceptual designation? Duchamp gleefully refused to publicly claim authorship, effectively complicating traditional legal notions of artistic ownership, and leaving the intellectual property status of his work intentionally ambiguous (Burke, 2018). Following Duchamp's bold gesture, later artists like Sherrie Levine (1947–) and Richard Prince (1949–) pushed the boundaries further by re-photographing established works and unapologetically presenting them as original creations. Levine's *After Walker Evans* (1981), for example, directly challenged assumptions about authorship, originality, and artistic merit, as she audaciously appropriated Evans's iconic photographs, prompting heated debates about artistic authenticity and the legal definitions of creative ownership (Karapapa, 2022).

Such provocative acts questioned whether authorship resides within the physical execution of a work or the intellectual vision underpinning it. This conceptual tension echoed earlier discussions within the Renaissance and Baroque periods regarding authorship and originality in printmaking and literature. By the twentieth century, however, this tension crystallized within the art world through conceptual artists who employed legal and bureaucratic means to define and protect their creations. Certificates of Authenticity, pioneered by conceptual artists Harold Cohen (1928–2016), Yves Klein (1928–1962), and others, transformed artistic authenticity from an aesthetic judgment into a legal and bureaucratic exercise. For instance, Cohen's generative program AARON produced algorithmic artwork, but it was Cohen's certification and authentication—his explicit authorship claims—that legitimized them within the art market (Mazzone & Elgammal, 2019). Similarly, Yves Klein's (1928–1962) certificates of authenticity for his immaterial artworks shifted authorship from tangible creations to documented intentions, effectively conflating legal provenance with creative legitimacy. Klein's work rendered physical manifestation irrelevant, insisting instead that it was the human creator's conceptual act that imbued a piece with artistic value, neatly encapsulated in the provocative practice of selling literally nothing but certificates to collectors who willingly accepted these invisible works as art.

Such certificates profoundly reshaped artistic authenticity, transforming it from a physical attribute to a documented intellectual declaration. As the conceptual art movement gained momentum, the reliance on certificates to authenticate artistic ownership became standard practice among contemporary artists and collectors. Legal recognition shifted to favor the idea itself rather than the artisanal labor traditionally required, thereby exposing significant shortcomings in copyright law, especially for nontraditional or performative works (Karapapa, 2022). This evolution underscored an essential shift: that originality was now more associated with the conceptual intervention of the artist than the craft of execution. Yet this innovative approach created tensions within traditional copyright law, which still struggled to accommodate nontangible artistic forms. While the courts continued emphasizing the necessity of originality—as seen in cases such as *Rogers v. Koons* (1992), where Jeff Koons was penalized for re-appropriating a photographer's composition—the growing prominence of conceptual art highlighted the law's limited capacity to recognize creativity not defined by manual labor (Nodelman, 2008).

Thus, historical precedents reveal a persistent negotiation between artistic innovation and copyright protections, demonstrating that while the medium may change, core tensions persist. Duchamp's subversion of artistic authorship and originality was not merely an avant-garde provocation; it was a challenge to the deeply entrenched perception that physical labor is the true measure of creative merit. His gesture, like those of Levine and Prince who followed, provocatively reoriented attention toward the significance of intellectual vision rather than manual craftsmanship, reminding us that creative legitimacy is less about how something is made and more about why it is made. At the same time, Duchamp's influence extended far beyond the realm of *readymades*, profoundly shaping contemporary debates about originality, artistic ownership, and creative authorship. His provocative works, particularly *Fountain*, continue to unsettle art historians, critics, and legal scholars alike, compelling a reassessment of traditional views on creativity and ownership. Conceptual artists, influenced by Duchamp, asserted that the value of art lay predominantly in the conceptual underpinning rather than the manual craft.

Despite such radical theoretical developments, the practicalities of the art market and established legal frameworks stubbornly cling to traditional notions of individual authorship, requiring clear attribution even when dealing with appropriation and conceptual art. The continued insistence on clearly defined authorship starkly contrasts the conceptual premise of appropriation art, evident in controversies surrounding artists like Jeff Koons (1955–), whose appropriative works regularly provoke legal battles—as in the notable *Rogers v. Koons* case (1992). There, Koons' deliberate replication of a photograph by Art Rogers was ruled infringement rather than transformative commentary, reinforcing copyright's emphasis on originality and tangible human creativity, rather than conceptual critique or commentary (Nodelman, 2008). Thus, while the art world may celebrate conceptual provocations, copyright law stubbornly remains tethered to a notion of material originality.

The tension over artistic authorship becomes more complex when considering collaborative or collective creative processes, challenging the romantic ideal of a singular artist as author. Large-scale artistic endeavors, from architecture and installation art to digital media projects, invariably involve multiple contributors, complicating straightforward claims of individual authorship (Cousijn, 2016). Graphic design theorist Michael Rock underscores this collaborative reality, arguing that most creative projects in contemporary settings resist neat authorship attribution. His influential assertion that "design is inherently collaborative" challenges the individualized authorship that copyright law traditionally requires, evoking a return to Renaissance workshop practices where collective production under the aegis of a master artist was commonplace. Despite this inherently collaborative tradition, copyright frameworks remain stubbornly fixated on singular, identifiable authors, creating friction in legally recognizing collective creative practices.

Amid this ongoing negotiation, the literary world found its own Duchampian counterpart in Kenneth Goldsmith (1961–), a key figure in conceptual writing who provocatively undermined conventional notions of authorship and originality. Goldsmith's conceptual projects, such as *Day*—in which he transcribed an entire issue of *The New York Times*—questioned literary originality by reducing authorship to the act of selection and transcription. His methods challenge traditional perceptions of literary labor and craftsmanship, echoing Duchamp's insistence that the conceptual idea behind a work holds greater artistic significance than the meticulousness of its execution. This radical approach democratizes writing, positioning every text as potentially ripe for creative repurposing, yet simultaneously reignites debates about intellectual property, authenticity, and creative integrity. Goldsmith's strategy thus echoes Duchamp's provocations, revealing the ongoing cultural tension between honoring originality and embracing the inherently transformative possibilities of recontextualization.

Poststructuralist theorists further deepen this challenge to authorial primacy, notably Roland Barthes (1915–1980), whose influential essay "The Death of the Author" (1967) famously declared the author's interpretive irrelevance. Barthes argued that the authority to define textual meaning no longer resides with authors but with readers, whose diverse interpretations construct meaning anew each time a text is engaged. Similarly, Jacques Derrida (1930–2004) deconstructed the long-held binary between speech and writing, exposing the inherent instability of meaning and revealing language as perpetually open to reinterpretation. Derrida's notion of textual instability fundamentally destabilized the idea of singular authorship, aligning literary theory with conceptual art's challenge to creative authority. Thus, poststructuralism complements conceptual art's subversion of traditional authorship by emphasizing fluidity and plurality in interpretation, further distancing creativity from singular, authoritative control.

The legal system has historically grappled—often unsuccessfully—with the slippery complexities of conceptual and appropriation art, precisely because traditional copyright frameworks emphasize tangible artifacts rather than ideas. *Fountain* (1917) by Duchamp typifies this predicament, challenging jurists and art historians alike: How can copyright law adequately protect a work whose value resides primarily in conceptual provocation rather than physical craftsmanship

(Bailey, 2017)? Just as nineteenth-century photography forced courts to recalibrate their notions of creativity—initially seen as mere mechanical reproduction—AI-generated art today unsettles definitions of originality, intent, and authorship. The same legal uncertainties that once plagued photographers now recur with generative art, prompting awkward judicial contortions to ascertain whether minimal human intent via prompt-writing constitutes sufficient authorship (Bailey, 2017). In effect, Duchamp's readymades continue to haunt copyright debates, persistently reminding observers that the crux of originality may reside in conception rather than manual execution, an irony not lost on today's lawmakers.

Exploring the pre-history of AI in art and literature reveals an enduring fascination with—and suspicion of—automated creativity. This narrative underscores an enduring ambivalence toward automation, reflective of humanity's simultaneous desire for innovation and anxiety over authenticity and control. Early examples from antiquity, such as mythic automatons or the programmable creations of Hero of Alexandria, embody this dual fascination and unease, foreshadowing current debates about human versus algorithmic authorship (Gerolemou, 2022; Moussa & Fekry, 2022). These early automata posed unsettling philosophical questions about the nature of creativity, begging us to ask what fundamentally differentiates human ingenuity from mere mechanical imitation. Modern artists and writers navigating generative technologies face remarkably analogous anxieties, as automation threatens traditional concepts of individuality and originality.

Such tensions reemerged vividly during the Renaissance and continued through the Enlightenment, where increasingly sophisticated automata—like Maillardet's poetic Draughtsman-Writer—embodied the era's enthusiasm for machines imitating human actions, alongside lingering skepticism about their artistic legitimacy (de Panafieu, 1984). This technological enthusiasm was matched by visionary thinkers like Ada Lovelace (1815–1852), whose collaborations with Charles Babbage suggested, presciently, that machines could create original works if suitably programmed, thus fundamentally reframing the relationship between creativity and human agency (Aiello, 2016). Lovelace's philosophical leap prefigured contemporary AI art's foundational concepts, notably the idea that automated systems can be creative partners rather than mere imitators. Yet the lingering question remained—and remains—where precisely to draw the line between human originality and mechanical replication.

By the mid-twentieth century, pioneers like Gordon Pask and Jean Tinguely further blurred the boundaries between machine imitation and creative autonomy through inventions such as MusiColour and interactive painting machines. These creations engaged dynamically with human input, producing unpredictable yet compelling artworks that provoked discussions about artistic intention and authorship—issues still central to contemporary debates on generative media (Pickering, 2007; Violand, 1990). The artistic experiments of this period directly challenged the conventional separation between human creativity and mechanized reproduction, anticipating today's complex negotiations over AI-generated art and literature. Thus, the pre-history of AI and copyright reveals an enduring tension between technological enthusiasm and cultural anxieties about authenticity and artistic control—a dialectic still evident in present controversies surrounding generative art. This legacy of conceptual challenge and technological experimentation in artistic and literary history underscores how contemporary anxieties about AI authorship are neither unprecedented nor unique. Instead, they represent a continuation of an age-old cultural ambivalence toward mechanization's role in creative fields, reinforcing questions that have long troubled artists, critics, and jurists: what defines genuine creativity, who owns it, and what role—if any—machines should play in artistic and literary production.

This chapter has demonstrated that contemporary debates about copyright, AI, and originality are neither novel nor unprecedented; rather, they reflect persistent tensions deeply embedded within cultural and legal traditions surrounding creativity and authorship. Historically, copyright protections evolved as responses to transformative technological disruptions—from Gutenberg's press and Renaissance printmaking privileges to modern-day generative tools—revealing society's enduring struggle to balance intellectual property rights with innovative technologies. Each period introduced unique challenges, yet the crux of these disputes remained remarkably consistent: the uneasy distinction between human and machine contributions, and the persistent question of whether creative authority resides primarily in tangible execution or conceptual intent.

This tension has been vividly illustrated across centuries and disciplines. From the skepticism of Baudelaire regarding photography's legitimacy as artistic authorship, resolved only by the U.S. Supreme Court's affirmation in *Burrow-Giles Lithographic Co. v. Sarony* (1884), to landmark literary cases such as *Wheaton v. Peters* (1834) and *Folsom v. Marsh* (1841), the historical record underscores copyright's enduring struggle with technological advancements. Furthermore, cases such as *Meshwerks v. Toyota* (2008) and *Thaler v. Perlmutter* (2023) emphasize that current legal frameworks are consistently challenged by AI-generated works. The persistent need for demonstrable human intervention and creative decision-making continues to shape the contours of intellectual property protections in an era defined by generative algorithms.

These historical precedents also provide important lessons for understanding contemporary anxieties about originality. Duchamp's provocative assertion of conceptual authorship, epitomized in *Fountain*, foregrounded a crucial philosophical pivot — from traditional craftsmanship toward intellectual and conceptual frameworks. Similarly, literary figures such as Kenneth Goldsmith and poststructuralist theorists, notably Roland Barthes, have further dismantled notions of fixed authorial intent, reinforcing that ideas can hold primacy over physical execution. These movements have consistently pushed boundaries, compelling artists, scholars, and legal professionals alike to reconsider the very nature of authorship, creativity, and intellectual property.

The legal history also shows a consistent evolution toward safeguarding authorship rights while adapting to cultural and technological shifts — from Albrecht Dürer's early struggle against unauthorized reproductions, through Renaissance privileges and Enlightenment-era print advancements, to contemporary rulings on AI-generated materials. Landmark cases such as *Folsom v. Marsh* (1841), *Feist v. Rural Telephone Service Co.* (1991), and recent rulings surrounding generative AI emphasize the necessity of original human contribution. At the same time, current lawsuits against platforms like Stability AI, OpenAI, and Meta highlight ongoing tensions around training data usage, authorship transparency, and ethical considerations.

The historical trajectory traced in this chapter highlights a cyclical pattern: Technological innovation initially elicits fears regarding the displacement of human creativity and identity, only to later become normalized within artistic and literary practice. The ambivalence toward AI thus emerges as part of a longstanding tradition of creative anxiety — each innovation feared, debated, eventually normalized, and then integrated into the broader fabric of human cultural expression. As AI-generated works continue to test legal and cultural boundaries, the enduring lesson remains clear: Creativity and originality have always been, and will continue to be, moving targets — shaped more by human perception and convention than by any fixed legal or technological standard.

References

Aiello, L. C. (2016). The multifaceted impact of Ada Lovelace in the digital age. *Artificial Intelligence, 235*, 58–62.

Aldridge, A. O. (1962). Benjamin Franklin and the "Pennsylvania Gazette." *Proceedings of the American Philosophical Society, 106*(1), 77–81.

Allen, W. (1986). Legal tests of photography-as-art: Sarony and others. *History of Photography, 10*(3), 221–228.

Armstrong, E. (2002). *Before copyright: The French book-privilege system 1498–1526.* Cambridge University Press.

Atilla, S. (2024). Dealing with AI-generated works: Lessons from the CDPA section 9 (3). *Journal of Intellectual Property Law and Practice, 19*(1), 43–54.

Bailey, B. (2017). Before, during, and beyond the Brillo Box: The impact of Pop on the 1964 edition of Duchamp's readymades. *Visual Resources, 34*(4), 347–363.

Barron, A. (2020). Copyright law and the claims of art. *SSRN.* https://doi.org/10.2139/ssrn.346361

Barthes, R. (2016). The death of the author. In *Readings in the theory of religion* (pp. 141–145). Routledge.

Bourne, R. (2015). *Lords of Fleet Street: The Harmsworth dynasty.* Routledge.

Brittain, B. (2024, September 26). Artist sues after US rejects copyright for AI-generated image. *Reuters.* https://www.reuters.com/legal/litigation/artist-sues-after-us-rejects-copyright-ai-generated-image-2024-09-26/

Burke, S. (2018). Copyright and conceptual art. In E. Bonadio & N. Lucchi (Eds.), *Non-conventional copyright* (pp. 44–61). Edward Elgar Publishing.

Burnett, F. H. (1887). *Little Lord Fauntleroy.* C. Scribner's sons.

Burylo, Y. (2022). AI generated works and copyright protection. *Entrepreneurship, Economy and Law, 3*, 7–13.

Busacca, A., & Monaca, M. A. (2025). Who's afraid of "Big Bad" generative AI? Brief notes on the IDPA provision against OpenAI ChatGPT. In D. Marino & M. Alberto Monaca (Eds.), *Generative Artificial Intelligence and fifth industrial revolution* (pp. 117–144). Springer.

Chiglintsev, E. A. (2015). Reception of the Icarus myth in the mass art of the late 20th-21st century. *Terra Sebus. Acta Musei Sabesiensis. Special Issue 2014. Russian Studies. From the Early Middle Ages to the Present Day, 2014*, 177–186.

Cohen, H. (1995). The further exploits of AARON, painter. *Stanford Humanities Review, 4*(2), 141–158.

Cooper, E. (2021). *Art and modern copyright: The contested image* (Vol. 47). Cambridge University Press.

Cousijn, M. (2016). Marcel Duchamp and the art of exhibition making. *Relief, 10*(2), 143–149.

Culver, M. (1984). An examination of the July 8, 1838 letter from Harriet Martineau to United States Supreme Court Justice Joseph Story as it pertains to United States Copyright Law. *Journal of the Copyright Society of the USA, 32*, 38.

Davis, D. A. (1991). *Feist Publications, Inc. v. Rural Telephone Service Co.*: Opening the door to information pirates. *Saint Louis University Law Journal, 36*, 439.

de Panafieu, C. W. (1984). Automata—A masculine utopia. In E. Mendelsohn & H. Nowotny (Eds.), *Nineteen eighty-four: Science between utopia and dystopia* (pp. 127–145). Springer.

Dimitrova, R. (2023). *Are AI-assisted works copyrightable?* Paper presented at the 2023 International Scientific Conference on Computer Science (COMSCI), IEEE. https://doi.org/10.1109/COMSCI59259.2023.10315917

Filson, L. (2017). Magic and mechanics: The late-renaissance automata of Francesco I de'Medici. In J. A. T. Lancaster & R. Raiswell (Eds.), *Evidence in the age of the new sciences.* Springer.

Fitzgerald, W. (1984). Aeneas, Daedalus and the labyrinth. *Arethusa, 17*(1), 51–65.

Frosio, G. (2024). Generative AI in court. In N. Koutras & N. Selvadurai (Eds.), *Recreating creativity, reinventing inventiveness* (pp. 3–44). Routledge.

Füssel, S. (2020). *Gutenberg and the impact of printing.* Routledge.

Geiger, C. (2023). Elaborating a human rights friendly copyright framework for generative AI. *International Review for Intellectual Property and Competition Law 2024, 55*(7), 1129–1165. https://doi.org/10.2139/ssrn.4634992

Gerolemou, M. (2022). *Technical automation in classical antiquity.* Bloomsbury Publishing.

Gervais, D. (2022, February 8). AI derivatives: The application to the derivative work right to literary and artistic productions of AI machines. *Seton Hall Law Review*, *53*, Vanderbilt Law Research Paper No. 22-12, *SSRN Electronic Journal*. https://doi.org/10.2139/ssrn.4022665

Greenhalgh, C. M. B. (2008). King, Pope, Emir and Caliph: Europe and the Islamic building boom. In *Marble past, monumental present* (pp. 327–361). Brill.

Growcoot, M. (2025, February 12). This is the first-ever AI image to be granted copyright protection. *Peta Pixel*. https://petapixel.com/2025/02/12/this-is-the-first-ever-ai-image-to-be-granted-copyright-protection-a-slice-of-american-cheese/

Hedrick, S. F. (2018). I think, therefore I create: Claiming copyright in the outputs of algorithms. *NYU Journal of Intellectual Property & Entrepreneurial Law*, *8*, 324.

Heilo, O. (2022). The ʿAbbāsids and the Byzantine Empire. In Maribel Fierro, M. Şükrü Hanioğlu, Renata Holod, & Florian Schwarz (Eds.). *Baghdād* (pp. 339–370). Brill.

Henderson, P., Li, X., Jurafsky, D., Hashimoto, T., Lemley, M., & Liang, P. (2023). Foundation models and fair use. *arXiv, abs/2303.15715*. https://doi.org/10.48550/arXiv.2303.15715

Horzyk, A. (2023, June 18–23). *How AI affects our understanding of musical works that should be protected by copyright*. Paper presented at the 2023 International Joint Conference on Neural Networks (IJCNN). IEEE. https://doi.org/10.1109/IJCNN54540.2023.10191524

Iaia, V. (2022). To be, or not to be…original under copyright law, that is (one of) the main questions concerning AI-produced works. *GRUR International*, *71*(9),793–812. https://doi.org/10.1093/grurint/ikac087

Ismantara, S., & Silalahi, W. (2025). The lawfulness of using copyrighted works for generative AI training: A case study of a US lawsuit against OpenAI and Perplexity AI. *JUSTISI*, *11*(1), 127–150.

James, T. B. (2024). Artificial Intelligence, copyright registration, and the rule of doubt. *Texas A&M Law Review Arguendo*, *12*, 1.

Joicey, N. (1993). A paperback guide to progress: Penguin Books 1935–c. 1951. *Twentieth Century British History*, *4*(1), 25–56.

Joyce, C. (2005). A curious chapter in the history of judicature: *Wheaton v. Peters* and the rest of the story (of copyright in the new republic). *Houston Law Review, 42*, 325.

Karapapa, S. (2022). Art and modern copyright: The contested image. *The Journal of Legal History, 43*, 116–117.

Kasap, A. (2021). Copyright and creative artificial intelligence (AI) systems: A twenty-first century approach to authorship of AI-generated works in the United States. Wake Forest. *Journal of Business & Intellectual Property Law, 19*, 335.

Krages, B. P. (2020). *The photographer's right.*

Leth, G. (1993). A protestant public sphere: The early European newspaper press. *Studies in Newspaper and Periodical History, 1*(1–2), 67–90.

Lucchi, N. (2024). ChatGPT: A case study on copyright challenges for generative artificial intelligence systems. *European Journal of Risk Regulation, 15*(3), 602–624.

Mathur, A. (2023, December 11). Case review: *Thaler v. Perlmutter* (2023). *Center for Art Law.* https://itsartlaw.org/2023/12/11/case-summary-and-review-thaler-v-perlmutter/

Mazzone, M., & Elgammal, A. (2019). Art, creativity, and the potential of artificial intelligence. *Arts, 8*(1), 26.

Mensah, G. B. (2023). Artificial intelligence and ethics: A comprehensive review of bias mitigation, transparency, and accountability in AI Systems. *Africa Journal for Regulatory Affairs, 10*, 1–27.

Moran, M. E. (2011). The history of robotic surgery. In Ashok K. Hemal and Mani Menon (Eds)., *Robotics in genitourinary surgery* (pp. 3–24). Springer.

Moussa, I., & Fekry, W. (2022). The influential impact and contributions of the scientific heritage of Mouseion's scholars towards renaissance and present-day technologies. *Journal of Tourism, Hotels and Heritage, 4*(1), 55–78.

Myers, G. (2021). Muddy Waters: Fair use implications of *Google LLC v. Oracle America, Inc. Northwestern Journal of Technology and Intellectual Property, 19*, 155.

Nadelmann, K. H. (1958). Henry Wheaton on American Law in the Jurist (London). *New York Law School Law Review, 4*, 59.

Newbery, J., Goldsmith, O., & Jones, G. (1765). *The history of little goody two-shoes.*

Nodelman, S. (2008). Hanging the work of art: Love and death in the Duchampian readymades. *Interstices: Journal of Architecture and Related Arts, 3*(1), 41–44.

Nordin, A. N., & Ramli, N. (2020). Regenerating Muslim inventors: The present future. *Ulum Islamiyyah, 31,* 1–18.

Patterson, L. (1997). *Folsom v. Marsh* and its Legacy. *Journal of Intellectual Property Law, 5,* 431.

Pickering, A. (2007). Ontological theatre Gordon Pask, cybernetics, and the arts. *Cybernetics & Human Knowing, 14*(4), 43–57.

Price, D., & Wells, L. (2021). Thinking about photography: Debates, historically and now. In M. Durden & J. Tormey (Eds.), *Photography* (pp. 11–82). Routledge.

Ren, L. L., & Zhang, L. (2024). To get-thy images: Comparing the fair use of copyright in AI machine learning in Singapore & UK. *Singapore Comparative Law Review 1*(1), 372.

Ren, R. (2023). Ahead of his time: Leonardo da Vinci's contributions to engineering. *Journal of Education, Humanities and Social Sciences, 21,* 18–25.

Rexer, R. (2019). Baudelaire's bodies, or redressing the wrongs of nude photography. *Word & Image, 35*(2), 126–140.

Samuelson, P. (2023). Generative AI meets copyright. *Science, 381,* 158–161. https://doi.org/10.1126/science.adi0656

Sancataldo, S. (2021). Art and Copyright: A Matter of Moral Rights. J. Art Crime, 26, 59.

Spica, E. (2024). Public interest, the true soul: Copyright's fair use doctrine and the use of copyrighted works to train generative AI tools. *Texas Intellectual Property Law Journal, 33*(1), 67–91.

Springmann, M. J. (2020). The Schlüsselfeld ship model of 1503. *The Mariner's Mirror, 106*(4), 390–407.

Turner, J. (Ed.). (2000). *The Grove Dictionary of Art: From renaissance to impressionism: Styles and movements in western art 1400–1900* (Vol. 1). Macmillan.

U.S. Copyright Office. (2025). *Copyright and artificial intelligence, Part 2: Copyrightability*. U.S. Copyright Office. https://www.copyright.gov/ai

Uzun, A., & Vatansever, F. (2008). Ismail al Jazari machines and new technologies. *Acta mechanica et automatica, 2*(3), 91–94.

Violand, H. E. (1990). *Jean Tinguely's kinetic art or a myth of the machine age* (Vols. I–III). New York University.

Wagh, S., Peerzada, D., & Rote, P. (2023). AI and copyright. *Tuijin Jishu/Journal of Propulsion Technology, 44*(3), 3431–3439. https://doi.org/10.52783/tjjpt.v44.i3.2053

Wan, Y., & Lu, H. (2021). Copyright protection for AI-generated outputs: The experience from China. *Computational Law & Security Review, 42*, 105581. https://doi.org/10.1016/J.CLSR.2021.105581

Yu, R. (2017). The machine author: What level of copyright protection is appropriate for fully independent computer-generated works? *University of Pennsylvania Law Review, 165*, 1245.

CHAPTER 3

Authorship Redefined

This chapter takes on the cherished yet historically dubious narrative that sole authorship has always dominated creative writing, starting with the reminder that even the Bible—sacred, revered, and authorially complicated — was often dictated by Apostles to anonymous slave scribes. This inconvenient detail punctures the myth of solitary genius that still haunts contemporary demands for transparency in artificial intelligence (AI)–assisted authorship. Navigating through medieval scriptoria, Renaissance ateliers, Enlightenment literary salons, and poststructuralist theorists dismantling notions of singular authorial control, the chapter underscores that literary production has long been a tangled, collaborative endeavor. Indeed, modern anxieties around AI-generated writing—often dismissed as "soulless" or "inauthentic"—echo historical discomfort with past technological disruptions yet overlook how authorship itself has always been an intricate dance of influences, collaborations, and ghostly intermediaries. This analysis advocates moving beyond rigid expectations of transparency to embrace a more historically and philosophically accurate understanding of creativity's inherently collective and delightfully ambiguous nature.

3.1 From Solitary Creation to Collaborative Innovation

In 1770, the famed *Mechanical Turk* (Figure 3.1)—an automaton bedecked in regal Ottoman attire—captivated Europe by convincingly playing chess, prompting astonished aristocrats to marvel at what appeared to be a thinking machine. Little did the amazed spectators know, a chess master secretly operated the contraption from within, embodying an early instance of human discomfort with opaque creative mechanisms (Stephens, 2023). Contemporary demands for transparency from generative authorship carry echoes of this anxiety: Suspicion thrives where creativity hides behind digital curtains. As society once fixated on clockwork

Figure 3.1: Joseph Racknitz, *The Turk*, 1789 (CC O)

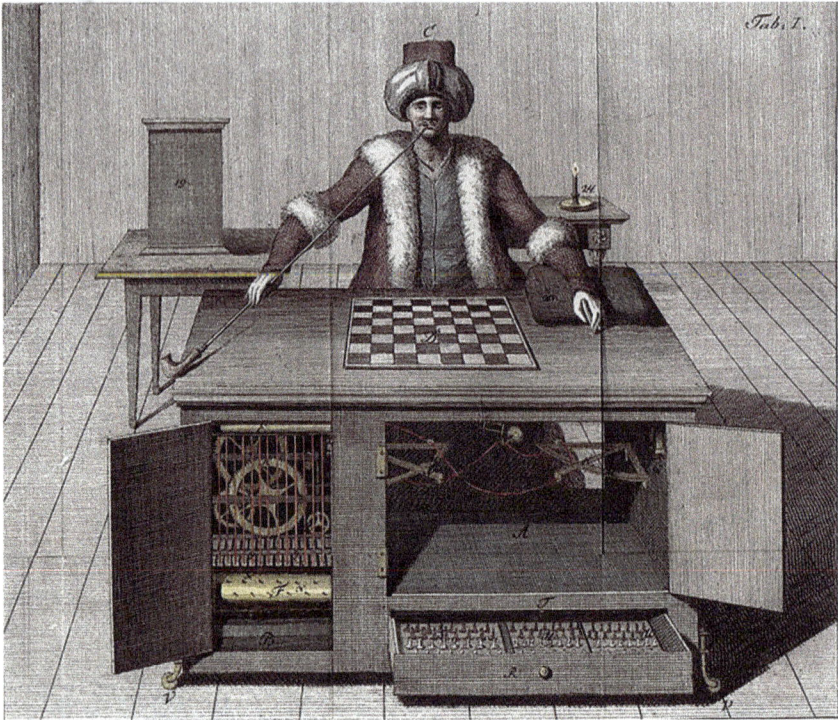

puppetry, now it fixates obsessively on algorithmic "black boxes," demanding detailed disclosures never historically imposed upon human authors (Perel & Elkin-Koren, 2017). Recall from Chapter 1 the writer and teacher David Perell, "The Writing Guy," who candidly confessed on Twitter/X in 2025 his existential unease triggered by generative writing tools' prowess: "This AI boom has set off an existential crisis in me…many of the skills I've developed and built my career on are becoming increasingly irrelevant." Despite excelling at instructing others to craft meticulous prose, Perell admits that carefully prompted large language models (LLMs) now outpace his best human effort—an unsettling revelation for professionals tethered to romantic notions of individual creative genius.

Yet Perell's discomfort is less a critique of machine capabilities and more a reflection of entrenched assumptions surrounding the role and identity of "the author." After all, the romantic conception of the solitary genius—locked away, scribbling feverishly by candlelight—is itself an historical construct, emerging relatively late in the literary timeline. Antiquity favored collective or dictated creation, illustrated vividly by biblical texts, crafted through oral transmission from apostles to anonymous slave scribes (Moss, 2024). The authorial ideal of individual genius is a Romantic invention, established well after authorship had already been enshrined legally by the Statute of Anne in 1710. This statute codified the notion of solitary authorship and laid the groundwork for modern copyright, yet masked centuries of collaborative and anonymous literary traditions (Cornish, 2010).

Moreover, Michel Foucault's incisive concept of the "author-function" sheds further light on the constructed nature of authorship. Foucault (1969) argues authorship serves as a cultural mechanism that privileges certain texts — literary works, for instance — by attaching authority to the individual's name, whereas texts whose value lies solely in replicability, like scientific results, often require no named author. Today's critics who challenge AI-assisted writing as a threat to authorial identity ignore this historic fluidity: What unsettles them is not just algorithmic involvement but the disruption of a carefully constructed authorial persona tied to personal authenticity, intellectual rigor, and creative labor. The discomfort expressed by figures like Perell thus reflects less a technological threat and more a confrontation with entrenched romantic notions about creative labor, which persist despite their relatively recent invention.

Further complicating this narrative is the shifting role of the audience itself. Traditionally relegated to the sidelines as passive consumers, readers now find themselves thrust into active participation, co-creating alongside AI-driven texts with every prompt or interaction. Historically, readers absorbed works passively — spectators, really, content to sit back and let authors perform intellectual acrobatics — but now audiences are becoming coauthors, remixing content in real-time dialogue with intelligent systems. Ironically, while contemporary calls for transparency seem to demand meticulous recordkeeping of every authorial sneeze, cough, and creative whim, the collaborative and iterative nature of modern generative processes makes such demands both impractical and laughably beside the point (Ilomäki et al., 2023). After all, meticulously documenting a generative text prompt session feels as absurd as transcribing every whispered prayer that guided medieval scribes' hands.

3.1.1 Deus Ex Machina: The Myth of Solitary Authorship

The notion of solitary authorship itself, which contemporary transparency demands cling to, is historically recent and rather peculiar. Authorship was traditionally collective, a group sport played across ancient civilizations from Mesopotamia to Early China, where texts emerged from community wisdom rather than individual genius. The concept of a singular, inspired author only emerged fully during Europe's medieval era and solidified legally in the eighteenth century, transforming writing into something attributable and legally protectable. Indeed, the one-paper-one-author model dominated academic publishing through the early twentieth century, neatly parceling scholarly credit. Yet even this solitary paradigm had its playful rebels: consider Daniel Defoe (1660–1731), who published under an astounding 198 pen names—a feat that makes today's transparency demands seem quaintly naïve by comparison (Defoe, 2022).

In fact, the historical practice of dictated authorship reveals that creating and interpreting texts has rarely been the solitary genius act romanticized today; rather, it has often been a collaborative affair. Long before AI-generated prose sparked existential panic in English departments, societies relied on scribes as mediators, capturing—and creatively embellishing—the authoritative pronouncements of prophets, rulers, or apostles (Schniedewind, 2024). Despite popular belief, sacred and philosophical texts weren't typically penned by their credited authors directly. Instead, like busy Egyptian viziers or harried Roman senators, many "authors" dictated to scribes who then took creative license, interpreting and embellishing as they transcribed (De la Durantaye, 2007). These scribes functioned as early ghostwriters, elaborating ideas with their own cultural or rhetorical flourishes, shaping messages to resonate with contemporary norms or ideological objectives (Rust, 2020). Consider the Apostle Paul (5–64/67 CE), a foundational figure in Christianity, who habitually employed scribes to record his theological missives, thereby co-constructing meaning in partnership rather than in isolation (Moss, 2024). Similarly, ancient Egyptian scribes or Roman record-keepers enriched official narratives, weaving interpretations into their recorded texts, thereby demonstrating that authorship has always been more of a collective enterprise than solitary genius. In modern times, generative tools like ChatGPT perpetuate this tradition, blurring distinctions between human intention and algorithmic improvisation—just digital scribes collaborating through computational creativity.

Revisiting dictated authorship thus offers intriguing insights into contemporary anxieties around generative AI, contextualizing these tools within a broader historical narrative of mediated creativity. Recent controversy around the impact on originality, intent, and authorship of generative tools mirrors longstanding practices, notably the scribal enhancements found within sacred writings like the Hebrew Bible and the New Testament (Mazzi, 2024). Such texts weren't static but vibrant, evolving documents whose meanings shifted as scribes inserted theological nuances and cultural contexts, not to mention the verbal commentaries from those who carried the manuscripts and offered interpretive responses to audience queries (Moss, 2024). This interplay of authorial intent, mediated expansions, and community interpretation parallels the modern interaction between humans and generative Natural Language Processing (NLP) tools like ChatGPT, whose algorithms creatively augment initial human prompts (Cain, 2024).

In both sacred and digital realms, the ambiguity surrounding authorship and intentionality generates vigorous philosophical and ethical debates. Critics of generative writing frequently question whether AI-produced texts should be considered genuine authorship or simply derivative reconstitutions of prior human-created works, akin to literary patchwork (Bukhari et al., 2024). Such discussions invoke the anxieties about originality and replication—an anxiety reminiscent of the unease surrounding early dictated texts, where authorship often involved unseen and uncredited intermediaries. The question remains: Whose intention truly shapes the final output—the original author, the scribe (or algorithm), or perhaps the audience interpreting the work (Coeckelbergh & Gunkel, 2024)? Consequently, societies face the necessity of redefining authorship to acknowledge collaborative, layered acts of creation. By connecting current AI-generated texts with historical practices of dictated authorship, it becomes clearer that creativity has never been a solitary affair but rather an intricate dance among originators, intermediaries, and interpreters. This perspective not only enriches understanding but also equips us to navigate the cultural shifts and challenges of the digital age.

Sacred texts like the Christian Gospels illustrate this intricate interplay among divine inspiration, human recorders, and interpretive communities, revealing how meaning is co-created rather than solely authored (Du Bois, 1987). Similar debates resurface today with synthetic media systems, prompting questions about the nature of generated texts as either meaningful collaboration or empty automated mimicry (Farina et al., 2024). Reception theory underscores how meaning-making involves active collaboration between the creator and the audience—a dynamic equally relevant to the interpretive communities surrounding generative AI outputs (Hutson & Harper-Nichols, 2023; Thompson, 1993; Virkler & Ayayo, 2023). Meanwhile, Foucault's notion of the Author-Function reframes authorship as a culturally and institutionally constructed role rather than an isolated act, illuminating parallels between ancient scribal intermediaries and contemporary human users mediating AI-generated texts (Canbul Yaroğlu, 2024; Foucault, 2003). These theoretical frameworks help reconceptualize authorship as fundamentally collaborative and historically fluid, challenging entrenched assumptions about originality, ownership, and meaning.

The historical practice of dictated authorship, rather than being a mere footnote in literary history, reveals a complex, collaborative tapestry woven deeply into the fabric of textual creation across ancient societies — particularly in the Mediterranean and Near East. Far from the romanticized image of a solitary author scribbling away in candlelit solitude, the authorship of texts in ancient Egypt was decidedly a team sport. Pharaohs and priests relied heavily on their viziers and trusted scribes to finesse religious decrees and doctrines, carefully shaping messages to reflect divine endorsement and maintain political cohesion (Meeks, 2003). Egyptian scribes (Figure 3.2) were not merely stenographers but coauthors, strategically tweaking wording to resonate with theological norms and ensure societal order — think ancient PR experts ensuring the gods were always seen as backing royal authority. Similarly, Roman scribes and orators functioned less as secretaries and more as political spin-doctors. Senators and influential statesmen depended on these scribes to polish speeches and legal texts, fine-tuning their rhetoric so it struck the perfect balance between eloquence and political expediency. Senators may have had the big ideas, but scribes transformed them into words that not only spoke — but sang (Dunn, 2004). The carefully edited pronouncements of senators, informed by the rhetorical skill of their scribes, showcased that ancient authorship was far from solitary; it was decidedly collaborative and inherently mediated, setting the stage for complex notions of shared authorship that reverberate today.

These ancient Mediterranean examples vividly illustrate that authorship across diverse societies was inherently a team endeavor, deeply collaborative, and embedded within broader cultural, theological, and political contexts. Whether in the royal courts of Egypt or the bustling forums of Rome, texts emerged not from solitary genius but from the negotiated, nuanced interplay between powerful figures and their trusted intermediaries. Such collaborative approaches challenge contemporary fantasies of singular authorship, revealing that messages have always been shaped as much by the scribe's interpretive hand as by the originator's visionary mind. This revelation helps contextualize the early Christian tradition, especially in the case of Apostle Paul, whose influential letters relied on collaborative authorship to navigate the complex theological and social landscapes of early Christianity (Barclay, 2018). Understanding these ancient partnerships reminds us that the current anxiety about collaborative creativity — whether with

Figure 3.2: Haremhab as a Scribe of the King, circa 1336–1323 BCE,
New Kingdom, 18th Dynasty (CC O)

human scribes or AI models—is neither new nor unprecedented. It is merely the latest manifestation of humanity's uneasy dance with delegated creativity.

In the realm of sacred texts, the Apostle Paul provides a compelling illustration of how dictated authorship served as a creative and collaborative endeavor. His letters, foundational documents for early Christian communities, represent far more than simple transcription; they reveal a nuanced partnership between Paul's theological vision and the interpretative flair of his scribes. While Paul dictated epistles filled with profound theological insight, scribes did far more than passively record—they actively engaged, reshaping rhetoric to ensure Paul's messages addressed pressing theological and social concerns of diverse early Christian communities (Eastman, 2021). Paul's epistles, particularly those addressed to communities such as the Corinthians, showcase this collaborative dance explicitly, with nuanced arguments carefully sculpted by scribal intervention. His letters weren't simply religious memos—they were carefully orchestrated theological documents, their rhetorical force honed by the scribes who added clarity and coherence, ensuring the letters would resonate powerfully across cultures and generations (Dunn, 2004). Far from merely jotting down dictated words verbatim, scribes crafted the theological complexity, coherence, and rhetorical flair that gave Paul's letters their lasting impact. Their contributions were neither trivial nor mechanical; they were essential intellectual partners who ensured messages aligned not only with their authors' intentions but also with the complex tapestry of local beliefs, customs, and rhetorical traditions. Recognizing the nuanced interplay between originators and intermediaries enriches our appreciation of historical authorship, especially when considering how analogous dynamics play out today with generative AI, whose outputs similarly hinge on interpretive human guidance.

The contributions of these scribes were indispensable yet deeply entangled in structures of systematic inequity and exploitation. Within ancient traditions such as Judaism, subordinate scribes regularly collaborated with prophetic figures, helping shape legal codes and visions into texts central to religious and cultural identity (Harper, 2012). Their interpretive labor was essential to the textual and doctrinal integrity of sacred writings, even if their social position denied them public acknowledgment. Such hierarchical relationships, while ethically problematic, illustrate how creativity and authority historically have intersected with power, status, and control—considerations still relevant as we debate the ethics of AI-driven authorship today (Freeman, 2023). Further complicating this narrative, dictated authorship in sacred texts frequently relied upon a broader network of participants—including those who carried and orally performed the texts—highlighting layers of interpretation beyond mere textual transcription. These ancient intermediaries enhanced the accessibility and relevance of sacred writings, adapting complex theological concepts into forms that diverse audiences could readily grasp (Harper, 2012).

The Hebrew Bible, with its metaphorical representations of divine qualities, provides another telling example. When passages referenced "the hand" or "face of God," these were rarely meant literally but rather served symbolic functions to bridge human and divine realities (Schoen, 1990). Scribes played a crucial role in refining such symbolism, crafting interpretations that addressed evolving theological concerns while preserving a sense of immediacy and cultural resonance (Harper, 2012; Schoen, 1990). Such historical dynamics echo intriguingly in the present-day interactions with generative AI: Humans craft initial prompts, which intelligent algorithms then interpretively expand into complex, contextually nuanced texts. These historical examples compel us to reconsider contemporary anxieties around AI-generated content. Far from a radical departure from tradition, these modern generative practices represent an evolution in the longstanding dialogue between authors, intermediaries, and audiences—a collaborative endeavor that has always been central to the creation and transmission of meaningful texts (Eastman, 2021, Harari, 2020, Rust, 2020). Today's debates over machine-generated writing strangely echo ancient dynamics. Algorithms act as modern scribes, invisibly interpreting prompts and producing sophisticated texts—yet their role is often minimized, reflecting lingering discomfort over collaborative authorship and attribution. Dictated authorship and modern generative writing share striking parallels, each marked by mediated creativity, interpretative elaboration, and layered meaning-making. In sacred traditions, scribes transformed divine dictation, embedding cultural nuances that ensured religious texts resonated across generations. Similarly, contemporary algorithms reinterpret human prompts, generating contextually enriched and adaptive content that extends beyond original intentions (Mazzi, 2024).

AI's role as intermediary in contemporary textual creation is frequently diminished, its outputs treated as suspiciously derivative, despite sophisticated interpretive processes (Coeckelbergh & Gunkel, 2024). Recognizing the historical continuity described above helps us question why we still cling so fiercely to solitary authorship—a concept historically anomalous and increasingly impractical in an era of mediated creativity. Indeed, acknowledging collaborative authorship enriches our understanding of textual production, positioning human and machine intermediaries as co-creators rather than invisible assistants (Wang et al., 2023). Perhaps embracing this collaborative heritage is precisely what the digital age demands, freeing us from outdated expectations of solitary genius and inviting more dynamic, inclusive models of creative agency. This examination thus illustrates the vibrant interplay between human originators, scribal mediators, algorithmic intermediaries, and active audiences. By recognizing these shared processes, we gain deeper insights into textual creation as inherently dynamic, evolving, and richly collaborative. These perspectives prompt us to reconsider traditional boundaries of originality and ownership, urging instead a model that appreciates how creativity emerges through mediation and iteration. From the scribes of antiquity to modern LLMs, mediated authorship is revealed as neither novel nor threatening—but profoundly human. In embracing this continuity, we can finally appreciate authorship as it truly is: an ongoing conversation across eras, participants, and, now, even across silicon processors.

In addition to theoretical complexities in distinguishing between individual and collaborative work (both with and without AI), there's another irony in today's panic over AI-generated hallucinations and inaccuracies. Critics fret about models like ChatGPT confidently asserting fictional references or fabricating scholarly citations—a legitimate concern, yet not entirely novel. After all, academia has long tolerated human error, bias, and outright invention under the guise of expertise. Yet, as Mollick's (2024) research elegantly illustrates, rather than eroding expert authority, generative AI has paradoxically enhanced it, reinforcing the importance of deep, specialized knowledge. While these algorithmic marvels indeed offer extraordinary potential to democratize information—providing instant expertise to novices—they have simultaneously raised the bar for human specialists, compelling them toward hyper-specialization. Indeed, as generative models improve, the barrier to outperforming them grows ever higher, ironically benefiting precisely those individuals already well-established in their fields. In other words, though anyone can now confidently chat about astrophysics or obscure Renaissance poetry, truly groundbreaking work—the stuff that pushes boundaries—will remain the province of hyper-specialists. Thus, generative AI, rather than making deep expertise obsolete, reinforces the importance (and perhaps elitism) of highly specialized domains, ensuring that those who hold detailed mastery will continue to shape how these tools are used, interpreted, and trusted.

3.2 The Shift in Authorial and Audience Roles

Writing and reading, those twin pillars of human communication, have historically served divergent purposes—though academia sometimes pretends otherwise.

On the one hand, a writer might meticulously detail the inner workings of a cell nucleus, distilling knowledge into clear, concise language; on the other, they might instead craft prose that delves into the nuanced turmoil of the human psyche, echoing the artist whose brush captures an emotional state rather than a textbook illustration. Similarly, readers approach texts with distinct motivations. Some pursue direct information—a transactional, efficient quest for facts—while others immerse themselves in stories to understand their own lived experiences through the mirror of another's imagination. As the literary snob Harold Bloom (1930–2019) lamented, the hallmark of humanistic study was precisely this immersive engagement with the multitude of human perspectives: By reading the thoughts, dreams, and anxieties of others, humans came to better understand their own place within society (Bloom, 2014). Yet the demands of a rapidly accelerating digital landscape now shift these traditional paradigms, prompting scholars and creatives alike to reconsider when the presence of a distinctly human voice remains essential, and when, perhaps, efficiency and practicality supersede nuanced expression.

Indeed, the act of writing—once imagined as an intimate dance between quill and parchment or fingers and keyboard—is increasingly shared with automated partners who churn out text faster than one can say "artificial intelligence." The humble scribe has evolved: From hieroglyph-etching Egyptian viziers and Roman senators dictating eloquent speeches to digital users inputting pithy prompts, the collaborative and interpretive roles in textual creation persist, albeit transformed (Rust, 2020). Cognitive engagement, traditionally heralded as the essence of writing, shifts from solitary reflection to mediated collaboration, reshaping how meaning is crafted and communicated in digital contexts. As algorithmic coauthors become commonplace, traditional distinctions between human and machine authorship dissolve, leaving us to grapple with uncomfortable yet essential questions: When do readers crave the unmistakable resonance of human emotion, and when is a quick algorithmic explanation perfectly sufficient?

3.2.1 The Evolving Dance of Human-Machine Authorship

Writing, at its core, is the craft of externalizing thoughts through symbols—letters, characters, and even emojis—to transcend boundaries of time and space. Merriam-Webster (2024) charmingly calls it both an action ("the act or process of one who writes") and a thing, which might reassure those worried about its existential crisis. Historically speaking, from clay tablets to digital screens, writing has provided humans with distinct channels of creativity, analysis, and, mercifully, a way to store grocery lists without memorizing them. Unlike the ephemeral nature of spoken language, written forms often create possibilities uniquely suited to nuanced, deliberate expression and reflection. Yet the real magic occurs when writing becomes a tool for externalizing our innermost thoughts, allowing us to refine them, dispute them, and sometimes marvel at their profound absurdity. This magic is amplified further in the digital age, as collaborative writing environments — particularly computer-mediated ones — reshape cognitive engagement, giving rise to novel opportunities for analysis, creativity, and self-expression.

Writing tools have evolved significantly from the earliest days of clay-tablet accounting in Mesopotamia to the hyper-sophisticated generative algorithms of today. While ancient civilizations primarily scratched out ledgers, modern writers now juggle handwriting, typing, and even AI-generated text that emerges as if summoned from thin air (or, more accurately, algorithmically conjured). These technological leaps don't just affect the means of writing — they reshape how we think as we write. For instance, contemporary research suggests collaborative digital environments, particularly in second-language learning contexts, lighten cognitive load and improve the quality of the writing produced (Leung et al., 2023). Such advances highlight the evolution of writing from mere manual documentation to a profound cognitive exercise, intertwined deeply with social and affective dimensions.

The process of writing has always involved more than mere transcription; it's a dialogue with oneself and others, mediated by symbols and tools that have evolved dramatically over centuries. Today, this dialogue increasingly includes algorithmic collaborators, demanding reconsideration of what writing — and authorship — truly mean. Where authors once held complete dominion over their manuscripts, carefully wielding quills or typewriters, they now find themselves guiding and curating content co-created with intelligent algorithms (Sarkar, 2023). This transformation highlights the shifting locus of human input from initial creation to editing and refining machine-generated content, complicating traditional notions of intellectual labor. The emergence of generative technologies, like LLMs, disrupts established norms, reshaping authorial roles into supervisory partnerships. Ironically, the meticulous labor once proudly displayed as a hallmark of intellectual rigor is now rivaled by — or even eclipsed through — the effortless generation of text by intelligent systems. Consequently, writing's authenticity is no longer gauged solely by the perspiration behind the prose but by the skillful interplay between human intention and computational ingenuity.

This digital scribal revolution further complicates established educational paradigms, demanding reconsideration of how writing itself should be taught, assessed, and understood. Historically, pedagogical practices emphasized solitary textual craftsmanship — students laboring carefully over each phrase, sentence, and paragraph. However, generative AI has rendered such isolated acts of textual production increasingly optional. Educators and scholars face the pressing need to develop frameworks that position these new tools as collaborative interlocutors, facilitating iterative ideation, structural refinement, and stylistic experimentation (Leung et al., 2023). This is not merely a practical shift — it heralds a fundamental reorientation of writing instruction, shifting the teacher's role from guardian of original content to curator of sophisticated machine-generated discourse. While purists may refuse to acknowledge this new variety of co-creative writing, the educational possibilities it unlocks are undeniable. The modern classroom thus becomes a place not only to practice solitary expression but also to cultivate the nuanced skill of steering algorithmic coauthors — a shift that promises both pedagogical innovation and existential anxiety in equal measure.

Since the beginning of the university, writing courses have claimed a lofty goal: enhancing students' critical thinking through clearly articulated ideas (Arapoff, 1967; Sinaga & Feranie, 2017). Writing assignments were long regarded as tangible evidence of intellectual engagement, reflecting how deeply students pondered, understood, or occasionally just panicked before deadlines (Ritchhart et al., 2011). Yet the meteoric rise of generative tools has reshuffled the pedagogical deck, blurring the once-clear lines between human-crafted prose and machine-assisted texts. With systems like ChatGPT at the fingertips of every undergraduate, writing has shifted from pure intellectual gymnastics to a collaborative choreography with algorithms. Naturally, this technological tango raises eyebrows among educators, who fear students might outsource analytical thinking altogether, reducing essays to mere prompt-engineered performances (Roe et al., 2023). But this anxiety may overlook opportunities for new instructional strategies: Perhaps, instead of diminishing analytical prowess, intelligent writing partners could amplify it — assuming students learn to co-create rather than merely copy-paste.

Today's increasingly ubiquitous generative platforms underscore the pragmatic (and ethical) ambiguities surrounding authorship and authenticity (Stojanovic et al., 2023; Tsao & Nogues, 2024). Since the advent of writing technologies — from typewriters to teleprompters — they faced similar initial skepticism, eventually becoming seamlessly woven into daily practice. Politicians confidently deliver teleprompted speeches ghostwritten by anonymous teams, yet few would question their authorship (Duffy & Winchell, 1989; Riley & Brown, 1996). Likewise, generative writing technologies now occupy a similar ambiguous space between assistance and authorship. The debate is not about to vanish; it's merely shifting, from arguments over the legitimacy of technology-assisted writing to distinctions along a continuum of human-to-machine engagement (Figure 3.3). English faculty like everyone else is being drawn slowly into acknowledging this new reality. The old systems of research, planning, composition, and revision are being replaced by prompts, curation, and iteration.

Figure 3.3: Types of Writing in the Age of AI

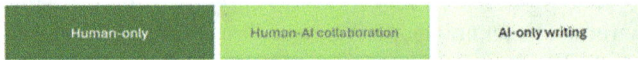

| Human-only | Human-AI collaboration | AI-only writing |

Consequently, the relationship between humans and automated writing tools is rapidly evolving into a sliding spectrum, with human-only writing at one end, fully automated generation at the other, and a vibrant zone of collaboration in the middle (Figure 3.3). In the human-driven process, intelligent systems merely support or refine preexisting human-crafted workflows. At the opposite extreme, writers become mere custodians of highly automated text-generating systems, their creative role reduced to curating outputs. The most engaging and innovative work will likely emerge from those who negotiate the middle ground, employing algorithms as interactive partners rather than passive servants or dominant masters (Caulfield, 2023). This necessitates a rethinking of traditional pedagogical models, pushing educators beyond the binary choice of accepting or rejecting algorithmic assistance. Instead, it compels scholars to help students cultivate skills that leverage computational creativity without relinquishing the essential spark of human ingenuity. After all, as writing and its methods evolve, so too must our understanding of what it truly means to write, author, and ultimately — think.

The emergence of digital text, including computer code and algorithmically generated content, must expand the scope of authorship and academic work. Textual outputs today are wildly varied, ranging from classical literature and contracts to synthetic prose spat out by algorithms — each no less valid in meeting the broad definition of persistent language representation (Giampieri, 2024). Since first putting pen to paper, the writer's role was unquestionably manual, labor-intensive, and solitary. Ancient scribes painstakingly etched clay tablets; later, pens and typewriters became symbolic extensions of human thought and creativity (Lyons, 2021). Yet these roles have steadily evolved, moving from manual inscription toward intellectual coordination and editorial oversight. Now, synthetic text tools such as ChatGPT or Claude elevate this transformation dramatically, shifting the writer's labor from composing every sentence by hand toward strategically prompting, editing, and refining sophisticated outputs produced collaboratively with intelligent software (Leung et al., 2023). These automated writing assistants blur distinctions between human and machine contributions, sparking debates about whether authorship is diminished or, paradoxically, enriched by algorithmic intervention. The result is a recalibration of authorship, redefining it less as a solitary endeavor and more as a dialogue between human creativity and computational prowess.

Consequently, what constitutes written work has radically diversified. The writing "product" now spans everything from classic literary narratives to automated business summaries, legal briefs, and even snippets of programming scripts — all comfortably nested within an expanded definition of textuality (Giampieri, 2024). The expanding capabilities of generative technologies disrupt conventional assumptions about creativity, originality, and labor; ironically, this disruption mirrors Socrates' ancient warnings, albeit updated for our digital anxieties. Thus, writing continues to serve as humanity's enduring elixir, not only of memory or knowledge but also of continuous reinvention in the face of transformative technologies.

3.2.2 Foucault and the Algorithmic Author-Function

The current redefinition of authorship through the lens of generative technologies elegantly dovetails with Michel Foucault's provocative notion of the "author-function." In 1969, Foucault proposed that the author's identity wasn't necessarily the secret sauce giving texts meaning; rather, the author serves as a function produced by societal expectations and interpretive communities. Fast forward to today, where sophisticated generative media solutions complicate authorship further. With machines contributing significantly — sometimes scandalously so — to text creation, the author-function no longer comfortably rests on purely human shoulders. Instead, it now sprawls across algorithms and digital intermediaries that shape not just how text circulates but also how it is conceptualized and received. Foucault would likely chuckle at seeing his theoretical grenade explode in the modern age of algorithmic authorship.

Foucault's critique dealt a hefty blow to romanticized myths of solitary, human-only creativity, shaking the foundations of what it means to author a text.

He undermined the once-untouchable ideal of human exceptionalism, opening doors for debates about the blurry boundary between creator and collaborator, human and machine. Intelligent video systems and NLP-powered tools like ChatGPT extend this challenge even further, forcing writers, scholars, and readers alike to wrestle with where human intent ends and automated authorship begins. Generative models compel a rethinking of creativity and originality, prompting essential questions: Does meaning lie exclusively within the human mind, or can it also reside within algorithmically produced interpretations?

Given this shifting landscape, it becomes clear that traditional distinctions between "human-written" and "machine-generated" content increasingly miss the point. Instead, the conversation might better focus on the collaborative interplay between humans and these synthetic intelligences — where humans provide the intent and context, and intelligent systems contribute efficiency, innovation, and surprising turns of phrase. Perhaps, rather than lamenting lost exclusivity, society should embrace this emerging collaborative mode of authorship, adapting legal and ethical frameworks to accommodate a partnership where human intention meets algorithmic capability. As generative writing continues to evolve, our definition of what it means to "write" must similarly adapt, potentially becoming more inclusive of both human ingenuity and machine creativity.

These transformations inevitably prompt reconsideration of writing pedagogy, particularly the conventional writing process taught within academic institutions. As outlined by Ferris and Hedgcock (2023), the traditional model of writing instruction involves stages (Table 3.1): beginning with prewriting — brainstorming, researching, and outlining — where ideas are conceived and structured. Next follows drafting, the moment ideas take shape into words, emphasizing development over perfection. Subsequent stages involve rigorous revision and editing, refining texts for clarity, coherence, and sophistication. In the past, these stages hinged upon individual cognition, reinforcing a belief in solitary authorship. Yet the introduction of generative writing platforms necessitates an updated pedagogy that embraces rather than resists collaborative creativity. Far from threatening writing skills, such technologies invite an enriched understanding of textual production, where humans actively collaborate with intelligent systems — embracing Foucault's assertion that authorship is indeed a dynamic and social construct (Ferris & Hedgcock, 2023).

Following drafting, comes the editing stage, a meticulous affair dedicated to refining grammar, punctuation, and stylistic finesse — essentially turning rough prose into something that won't make a grammarian cringe. Finally, the polished piece reaches its audience through publication or formal submission, marking

Table 3.1: Traditional Writing Process

Stage of Writing Process	Description	Activities Involved	Primary Objectives
Conceptualization and Planning	Generating initial ideas, organizing thoughts, and preparing outlines to clarify the writing's intent and target audience.	Brainstorming ideas, conducting preliminary research, and drafting outlines.	Clarify concepts, identify purpose, and understand audience needs.
Drafting	Composing the initial version of the text, emphasizing the articulation of ideas without excessive concern for precision or final quality.	Writing initial paragraphs, expanding main arguments, and elaborating on outlined ideas.	Establish content structure, develop primary arguments.
Revising	Evaluating and restructuring the draft to improve overall clarity, logic, and coherence.	Assessing arguments, reorganizing content, refining clarity and structure.	Enhance readability, coherence, and logical progression of ideas.
Editing	Refining textual details, focusing on grammar, punctuation, style, and linguistic accuracy.	Proofreading text, correcting errors, and improving stylistic elements.	Ensure grammatical accuracy, linguistic refinement, and textual polish.
Publishing	Completing and presenting the finalized text for feedback or dissemination to the intended audience.	Finalizing and submitting the work, sharing the text publicly or with evaluators.	Communicate effectively, present a polished final product.

the culmination of earlier stages into a cohesive, carefully crafted whole. Integral to this iterative journey are feedback loops, particularly cherished in higher education, where peer reviews and instructor critiques form a perpetual cycle of revision and enhancement. Writing, in this sense, is emphatically not linear but rather delightfully recursive, looping back repeatedly through stages until clarity triumphs over initial chaos. As rhetorical considerations—audience, purpose, and genre—take center stage in contemporary pedagogy, the writer is constantly nudged to shape and reshape content with an eye toward effective communication, not merely grammatical correctness.

Throughout this multistage endeavor, the roles of author and instructor ebb and flow, dynamically reshaping their interplay and the resulting sense of authorship (Myhill et al., 2023). At the prewriting phase, authors take center stage, independently exploring and defining the core ideas, intentions, and target audience for their work. Drafting further emphasizes authors' centrality as they articulate their vision into structured text, even as instructors gently advise on effective approaches. Yet, as the revising and editing stages unfold, authors enter a critical dialogue with instructor feedback, navigating between external insights on coherence and clarity and their own conceptual integrity. Here, authors balance external critique with personal voice, preserving authenticity while accommodating necessary revisions. Ultimately, the publishing phase solidifies this duality, as authors integrate editorial guidance while asserting their creative ownership in the polished, finalized work. Authorship thus emerges as a shared endeavor, woven from the interplay between external guidance and individual creative agency.

When generative technologies like LLMs enter the authorial workflow, the traditional writing process transforms significantly, emphasizing augmentation, co-intelligence, and collaborative creativity (Table 3.2). Rather than single-handedly managing every stage from drafting through revisions, authors now work symbiotically with intelligent writing tools that actively assist with ideation, initial composition, and editorial refinement. These technologies swiftly produce drafts based on targeted prompts, suggest improvements to grammar and style in real-time, and offer a constant stream of instant feedback—like having a particularly opinionated editor always peering over one's shoulder. Such co-authorship repositions the human author from sole creator to orchestrator, overseeing and guiding computationally produced drafts toward refined, coherent final texts.

This shift simultaneously redefines the author's role and complicates notions of authorship. Whereas traditional authors assumed full ownership over both the drafting process and the end product, the integration of synthetic media tools introduces a collaborative dimension that challenges conventional assumptions about originality and intellectual labor. Authors must now cultivate a sophisticated editorial discernment, carefully evaluating AI-generated content to ensure accuracy, authenticity, and alignment with their original intent. Rather than diminishing human creativity, this collaborative arrangement foregrounds

Table 3.2: AI-Augmented Writing Process

Stage	Author's Role	AI's Role	Instructor's Role
Prewriting	Explores ideas, grapples with concepts, and outlines content.	Suggests directions, offers thematic inspiration, and generates initial ideas based on prompts.	Guides strategic thinking, nudging authors toward clarity in purpose, audience, and genre.
Drafting	Translates ideas into initial written form, constructing arguments.	Crafts initial drafts from prompts, recommends phrasing, and helps bridge gaps in content.	Advises on balancing algorithmic assistance with preserving the author's unique voice, offering feedback on tool usage.
Revising	Evaluates drafts critically for coherence, flow, and overall effectiveness.	Suggests structural improvements, alternative sentence constructions, and content refinements.	Provides higher-level insights into argumentation and coherence, emphasizing meaningful revisions beyond superficial changes suggested algorithmically.
Editing	Fine-tunes language, correcting grammar, style, punctuation, and formatting.	Offers immediate, detailed corrections to grammar, style, punctuation, and formatting consistency.	Oversees accuracy and guides responsible application of intelligent editing tools, emphasizing final polish and precision.
Publishing	Prepares and submits the final refined document.	Finalizes document format, ensuring presentation quality through consistent stylistic adjustments.	Confirms quality, accuracy, and appropriateness of the final work, facilitating polished presentation.
Feedback	Integrates feedback for iterative improvement of future writing.	Provides rapid, continual feedback on structural coherence, style choices, and grammatical correctness.	Supplies comprehensive, contextual feedback to foster enhanced critical analysis and intellectual growth.

authorial decision-making, highlighting the author's expertise in curating, critiquing, and refining outputs. Thus, the contemporary author navigates a nuanced partnership with intelligent algorithms—transforming writing into an iterative, dynamic exchange that questions and expands conventional notions of what it means to craft text.

In this evolving authorial landscape, the distinctions illustrated in Figure 3.3 gain greater nuance through the elaborations presented in Table 3.3. Human-only writing remains the bastion of traditionalists, where every word penned, each argument

Table 3.3: Alignment of Human-AI Collaboration in Writing

Stage	Author's Contribution	AI Contribution	Instructor's Contribution
Human-Only	The author independently generates concepts, develops arguments, drafts content, and performs all revisions, managing the entire creative process without computational tools.	None; all idea generation, structuring, and drafting rely exclusively on human cognitive and creative efforts.	Provides guidance, feedback, and evaluation based entirely on human-generated content.
Human–AI Collaboration	The author initiates content through prompts and selects, edits, refines, and integrates algorithmic suggestions into their overall vision.	Provides initial drafts, content suggestions, structural enhancements, and stylistic refinements based on prompts and iterative interactions.	Supports authors in critically evaluating and effectively integrating algorithm-generated content, ensuring original intent and voice are maintained.
AI-Only Writing	The author's role is limited to providing basic initial instructions or minimal oversight, with the algorithm handling the content creation process autonomously.	Autonomously generates content, structures ideas, and applies stylistic and creative choices with minimal human oversight.	Ensures the final product meets academic, ethical, and professional standards, emphasizing the importance of transparency and accuracy.

constructed, and every structural nuance are entirely reliant on human intellect and creativity. Here, authors carry sole responsibility for both the creative spark and intellectual rigor, producing works that reflect personal insight, meticulous labor, and distinctive style—traits cherished, yet painstakingly cultivated. In contrast, the collaborative mode blurs these neatly drawn boundaries, introducing sophisticated tools as active partners. Generative systems like ChatGPT or Claude step in with initial drafts or narrative suggestions, which human authors refine, embellish, or reshape. Structural coherence similarly emerges through this interactive dynamic, with intelligent systems proposing frameworks or clarifying argumentation, and authors refining these suggestions to suit rhetorical goals. Creativity itself becomes a conversation between the algorithm and the human mind (and hand). Crucially, ethical accountability remains with the author, necessitating transparency about how extensively intelligent assistants influenced the final work.

At the far end of this spectrum lies automated authorship, wherein generative algorithms take center stage, composing content, organizing ideas, and proposing creative solutions autonomously. Human involvement in this scenario dwindles to minimal oversight, typically confined to confirming factual accuracy or stylistic appropriateness. Yet, even here, human supervisors bear ethical and intellectual responsibility, maintaining accountability for outputs that must meet scholarly, professional, or ethical standards. Ironically, while these tools promise to alleviate human labor, they simultaneously heighten the importance of human vigilance in ensuring accuracy, reliability, and ethical compliance. As writers navigate these evolving roles—shifting from sole creator to editor or even curator—the enduring responsibility for textual integrity underscores the indispensable role human judgment continues to play, despite the expansive capabilities of algorithmic coauthors.

The integration of generative tools into academic writing is reshaping perceptions of authorship within scholarly contexts. Academic publishers have begun formulating clear guidelines to manage the role of these generative assistants in the composition process. Notably, journals like *Nature* and publishers such as Taylor & Francis emphasize that although generative software can aid in drafting or refining manuscripts, ultimate accountability for content remains firmly with the human author. Any use of such technologies must be transparently disclosed, typically within methodology sections or acknowledgments, depending on their level of contribution (Ganjavi et al., 2024). This cautious stance reflects an interim phase in academic practice, where the involvement of intelligent software is acknowledged but meticulously regulated to safeguard the perceived value of human intellectual endeavor. For instance, publishers now clarify that employing generative tools for tasks such as summarizing data or structuring findings necessitates explicit disclosure, recognizing their potential impact on scholarly interpretation. This distinction is particularly evident when comparing disciplines: While fields such as the humanities continue to prioritize human creativity and interpretative nuance, areas like scientific research increasingly accept collaborative machine input. The divergence underscores academia's enduring emphasis on human agency and critical thought, especially in disciplines reliant upon subjective insights and interpretative rigor.

Academic publishers are progressively recognizing the dynamic roles that generative writing platforms and human authors play within scholarly composition (Hosseini et al., 2023). As intelligent text-generation software continues to advance, publishers are forced to delineate precisely the boundaries of acceptable use. Although human writers retain ultimate accountability for the final content, publishers insist upon clear disclosures of how generative software contributed, thereby ensuring transparency and preserving scholarly integrity. This evolving human–machine relationship underscores the necessity for refined frameworks to accurately assess these collaborative dynamics. As generative technologies become more sophisticated and deeply integrated into academic practices, scholars and institutions must develop clearer models for categorizing the contributions of synthetic media solutions and determining how to assign authorial credit. These frameworks must account not only for the growing capabilities of intelligent systems but also for the enduring necessity of human interpretive oversight. Establishing such nuanced guidelines will enable academia to move beyond its current transitional stage and accept generative platforms as integral collaborators in knowledge creation.

Theorizing a more nuanced framework for generative-assisted writing (Figure 3.4) necessitates moving beyond the simplistic, linear models dominating current discourse. Comparable to contemporary understandings of neurodiversity—which now recognize the interplay between multiple cognitive conditions—the collaboration between human authors and generative systems should be envisioned as multidimensional rather than unidirectional. Traditional frameworks, which simplify the interaction into a linear relationship, inadequately capture the nuanced partnership between human creativity and algorithmic suggestion throughout the writing process. In a multidimensional approach, each axis represents distinct components of textual production, reflecting varying degrees of human and synthetic collaboration.

Figure 3.4: Multidimensional Model for Human–AI Collaboration in Writing

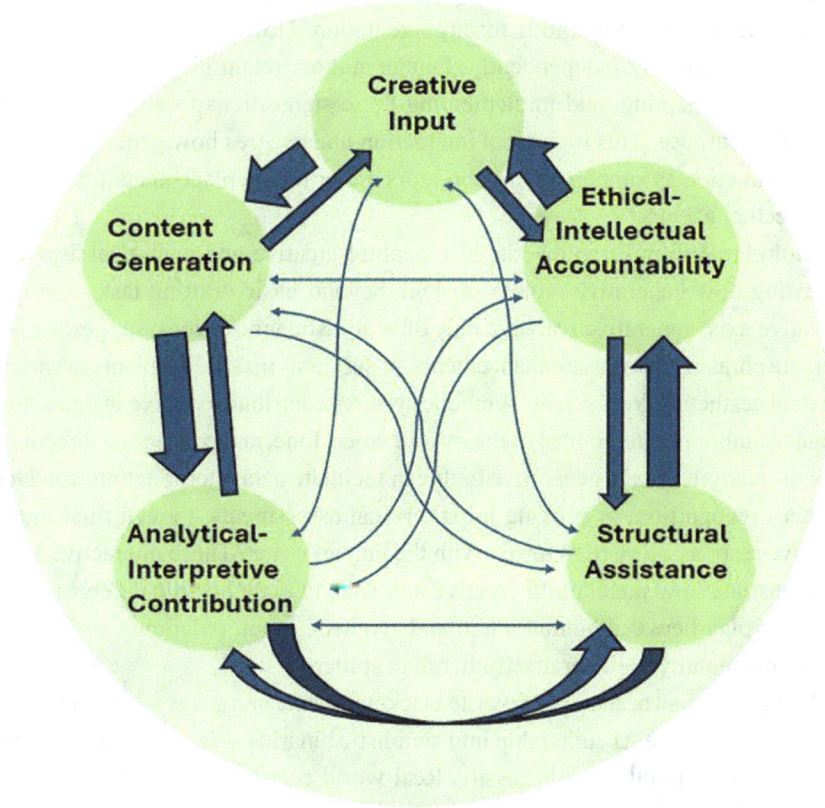

For example, on the content generation axis, generative platforms may offer assistance ranging from simple idea prompts or conceptual suggestions to generating extensive draft sections of text. This flexibility mirrors the overlapping traits observed in neurodiversity, where multiple cognitive factors combine uniquely within each individual. Yet, despite the robust capabilities of generative software, the human author remains critical as the final arbiter, selectively incorporating algorithmic suggestions to achieve the intended intellectual outcome. Such collaboration significantly revises the traditional image of the author as a solitary creator, promoting instead a more fluid and integrated concept of shared authorship.

On the structural assistance axis, the collaborative dynamic is further emphasized. Just as environmental conditions shape the expression of neurodiverse traits, generative assistants offer structural guidance by recommending organizational patterns or methods for ensuring textual coherence. Intelligent writing tools can highlight structural inconsistencies, propose alternative arrangements of content, or suggest enhanced methods for argumentation. However, these systems do not operate entirely independently; human authors retain the essential role of evaluating, adapting, and implementing these suggestions to align with their intended purpose. This reciprocal interaction underscores how generative platforms function as supportive collaborators rather than replacements for human intellectual agency.

Multidimensional frameworks also capture creative and analytical aspects, showing how generative tools contribute beyond basic drafting tasks. On the creative axis, generative software may offer stylistic innovations, suggesting alternate phrasing, narrative enhancements, or subtle stylistic adjustments to enrich textual aesthetics. Yet, even as synthetic systems contribute creative insights, the human author retains control of the overall voice, tone, and conceptual direction. On the analytical axis, generative tools can facilitate data interpretation, conduct pattern recognition, or provide initial critical assessments, though final interpretive responsibility rests firmly with the human writer. These interactive axes demonstrate how meaningful creative and analytical authorship emerges from the interplay between human intent and synthetic input, redefining authorship as fundamentally collaborative rather than solitary.

Moving beyond academia's favorite black-and-white narratives—which conveniently pigeonhole AI authorship into simplistic binaries—is not just a scholarly tweak; it's an intellectual necessity. Real-world collaboration with intelligent systems is messy, nuanced, and complicated, much like a great dinner party conversation—everyone contributes, the insights overlap, and you leave slightly unsure who exactly deserves credit for that brilliant final thought. Embracing this complexity offers a richer, more accurate picture of the creative process, turning us from rigid gatekeepers into savvy navigators of shared authorship.

3.2.3 The Active Audience and Changing Purpose of Reading

Just as our understanding of authorship has matured, so too has our notion of the audience. Previously, audiences approached texts and artworks equipped with distinct expectations shaped by prior experiences, linguistic fluency, and familiarity with visual and textual conventions. They looked upon artists and authors as singular visionaries who bestowed meaning upon their creations. Within this framework, the audience was expected to passively absorb the author's intended message, seeing the world through their eyes and marveling at their exceptional insight. Yet with the rise of postmodern and poststructuralist theories, this once-sturdy hierarchy has been flipped on its head. The audience is no longer merely a passive recipient but an active participant, co-creating meaning in a fluid, inter-pretive dance where the authority of the author fades, replaced by the dynamic interplay between reader, text, and context. Or to put it more accurately perhaps, the arguments made by postmodern theorists and entertained by academics for decades as theoretical daydreams or sites of political activism have now shown up as simple facts of the matter. The author-function really has become a function.

The purpose of education has historically evolved in response to shifting societal demands, with reading and writing consistently at its core as fundamental skills adapted for academic, professional, and cultural contexts. The rise of sophisticated generative AI tools, such as ChatGPT, has prompted profound reconsiderations about the ongoing relevance and nature of traditional literacy skills. These gen-erative technologies challenge longstanding conventions, shifting the primary educational goal from producing and consuming texts toward critically evaluat-ing, synthesizing, and effectively applying content generated through machine intelligence. This shift underscores a significant debate: Should educational and professional sectors prioritize cultivating deeper critical thinking and ethical frameworks suited for AI-mediated environments, or should emphasis remain on enhancing the technical skills required to effectively interact with and manage these innovative technologies? Such questions invite further exploration into how reading and writing have transformed under the influence of generative AI, reflect-ing broader cultural shifts in information engagement and intellectual practice.

Up until now, reading has served diverse purposes — academic, professional, and leisure — each shaped uniquely by technological and societal conditions (Nicholson, 2012). Although this diversity persists, advanced generative tech-nologies have substantially altered how individuals across these domains interact with texts. Previously, traditional literacy emphasized deep, linear engagement with static texts, prioritizing comprehension, analysis, and retention. The con-temporary environment, in contrast, increasingly values the ability to critically navigate and assess dynamic, synthesized content generated instantaneously by advanced language models (More, 2023). Such shifts necessitate reconsidering not only the skills required to read effectively but also the purposes and expec-tations surrounding textual interaction itself (George, 2023).

In scholarly contexts, reading practices have traditionally focused on meticulous analysis, comprehension, and the deep exploration of established literature. Today, however, generative tools like ChatGPT are capable of rapidly producing summaries, analytical frameworks, and even preliminary research reviews, fundamentally changing scholarly engagement with texts. Consequently, readers are increasingly required to redirect their cognitive resources from textual consumption to assessing the reliability, accuracy, and validity of these algorithmically derived outputs (Nauta et al., 2023). Researchers employing AI-driven systematic reviews, for instance, must maintain vigilance to ensure that synthesized outputs meet rigorous scholarly standards and avoid oversimplification or inadvertent bias (Castillo-Segura et al., 2023). This trend necessitates a heightened level of digital and critical literacy, highlighting the evolving competencies essential to contemporary scholarship (Ilomäki et al., 2023).

Professional reading practices have similarly shifted from static engagement with lengthy reports toward agile assimilation of machine-generated insights. Tools employing generative algorithms have become instrumental in managing the increasing volume and complexity of professional data streams. For example, intelligent systems in sectors such as information technology, business management, and healthcare now swiftly summarize information, enabling professionals to distill critical insights more efficiently. However, with the heightened capacity for instant content generation comes the responsibility to rigorously evaluate generated data, necessitating advanced interpretive skills and professional judgment (Noy & Zhang, 2023). This underscores the emerging importance of proficiency in strategically utilizing and interpreting synthesized information in high-stakes professional contexts.

Leisure reading has also experienced transformation through personalized, AI-curated content recommendations. Platforms such as Kindle and Goodreads leverage intelligent algorithms to refine and tailor reading suggestions, influencing reader preferences and engagement patterns. While this personalization increases access to relevant and appealing content, it simultaneously risks limiting exposure to algorithmically less favored texts or genres, inadvertently constraining cultural and intellectual exploration (Torppa et al., 2020). Furthermore, generative technologies increasingly facilitate personalized, dynamic narratives that invite readers to participate in co-creating content, thus blurring traditional distinctions between readers and authors (Fathoni, 2023).

As such, new competencies in reading are required to effectively engage with texts in the current intellectual landscape. Rather than emphasizing conventional literacy, proficiency now involves critically assessing AI-generated summaries, analyzing synthesized datasets, and adeptly navigating personalized content recommendations. Recent research argues that educational institutions must incorporate training in digital literacy and critical evaluation, equipping individuals to more effectively interact with dynamic information environments (Preiksaitis & Rose, 2023). Although the particular competencies emphasized may vary across contexts and disciplines, the broader shift toward sophisticated engagement with AI-mediated information marks a fundamental transformation in how society conceptualizes literacy.

3.2.4 From Execution to Ideation

Moreover, although earlier iterations of generative AI tools were prone to frequent inaccuracies or "hallucinations," significant improvements in recent versions — enabled through real-time connections to academic databases and authoritative online sources — have drastically enhanced reliability. Misinformation generated by these advanced systems has been reduced substantially, now reportedly below 2%, a milestone achievement in ensuring trustworthy outputs (Preiksaitis & Rose, 2023). Consequently, rather than diminishing the need for robust literacy, this technological evolution has redefined its parameters, emphasizing competencies in critical evaluation, strategic synthesis, and effective integration of AI-curated content across personal, professional, and academic domains.

The idea itself is now paramount. Today's human contribution in writing centers on ideation, curation, and editing, and not on the execution. This marks a profound departure from how intellectual property, copyright, and scholarly creation have been traditionally understood for centuries. As explored in Chapter 2, copyright law evolved explicitly to protect tangible expressions, emphasizing the physical act of creating content. Yet, as automated generative tools increasingly handle the "execution," the legal emphasis on tangible expression appears increasingly obsolete. This shift also poses intriguing implications for readers, viewers, and audiences: If ideas are the real currency of creativity, copyright frameworks — which historically neglect idea protection — risk losing relevance, calling into question mandatory AI disclosure practices.

This line of thinking resonates strongly with Amjad Masad, CEO of Replit, who boldly challenges the Silicon Valley truism that execution eclipses ideation. In a provocative interview with Matt Turck on The MAD Podcast (February 6, 2025), Masad proposes that the primary value in creative fields is shifting from execution toward ideation, problem-spotting, and conceptual innovation. His core premise is that generative technologies, such as Replit's coding assistant, increasingly automate the labor-intensive tasks of coding, debugging, and design. Consequently, the once-prized execution skills — like flawless coding or perfect grammar — become less decisive factors in success. Instead, the true bottleneck becomes the originality and value of ideas. In Masad's words, "finding problems to solve is a skill" (Turck, 2025, 1:13:17), underscoring a significant shift toward intellectual creativity over technical execution. Historically, Silicon Valley wisdom celebrated execution over ideas, yet the rise of generative tools reverses this hierarchy, ironically suggesting that "ideas," often dismissed as plentiful, may now become the more scarce and valuable asset.

Masad's insight further intersects with ongoing discussions about content credibility in a landscape increasingly populated by generative systems. On the Techmeme Ride Home podcast (2025), speakers emphasized the vital human role in curating and editing AI-generated outputs. They introduced the concept of "slop" — AI-generated content created without human oversight — as opposed to curated, credible work where a human explicitly "attaches their credibility" and reputation (40:54). The speakers humorously but pointedly suggest that without human curation, AI content resembles empty calories — "slop" — filling but devoid of meaningful nourishment. They emphasize a hybrid model, where generative systems provide diverse variations, but the human author serves as the essential "Last Mile" arbiter (41:46), verifying accuracy and stamping content with credibility. Credibility, therefore, rests on accountability — someone must "attach their name" to the content, signaling reliability and intentionality that AI alone inherently lacks (42:05). This reassertion of the human curator's centrality underscores a new equilibrium where human intellectual responsibility balances the efficiency and breadth offered by generative tools.

Such developments foreground critical questions about intellectual property and authorship in an increasingly automated environment. Current copyright frameworks prioritize tangible expression and concrete execution — elements increasingly outsourced to algorithms and generative engines. As execution becomes progressively automated, intellectual property laws rooted in protecting tangible, manually produced outputs must adapt. The value of intellectual work may soon rest more heavily upon ideation, conceptual originality, and human oversight — elements traditionally unprotected by copyright. This evolving landscape suggests that traditional legal paradigms may soon need reconsideration, shifting from protecting execution to incentivizing and safeguarding the conceptual ingenuity behind creations.

In this emerging framework, the value of human contribution in writing and creation lies primarily in ideation, curating, and editing rather than execution.

Masad asserts boldly that what matters is the selection, concept, and intellectual rigor — not the mechanical creation itself. In this respect, both historical and contemporary shifts underscore the need to reevaluate how society understands and protects creative output. By acknowledging ideation's centrality and adapting legal and educational frameworks accordingly, we recognize a landscape where human ingenuity and algorithmic efficiency blend seamlessly, creating a richer, more nuanced understanding of authorship in the age of generative tools.

One last digression: When the Venetian Senate granted Johannes of Speyer exclusive printing rights for five years in Venice and its dominions in 1469, skeptics warned of imminent cultural catastrophe, fearing mass-produced books would strip scholarly works of their perceived authenticity. They lamented that true wisdom would suffer if ideas could be so effortlessly replicated and disseminated. Sound familiar? Today's insistence on transparency in generative media — demanding that authors meticulously disclose every algorithmic assist — is similarly driven more by anxieties about control than by genuine ethical necessity. Indeed, such mandates are historically unprecedented: Artists and writers have always relied on hidden scaffolding, from scribes and assistants to editors and ghostwriters, without being obliged to provide a detailed audit trail of every creative decision. As the analysis throughout this chapter demonstrates, creativity thrives precisely because of its inherent ambiguity and partial concealment; demanding absolute transparency is not only unrealistic but antithetical to the creative spirit itself. Ironically, while society pushes for full disclosure, the truly innovative work has historically emerged from processes where the precise origins remain artfully obscured. Therefore, navigating authorship in the age of generative assistants requires abandoning rigid, solitary ideals, and embracing instead a more nuanced, collaborative vision of creation. In doing so, we might recognize that authenticity lies less in documenting every computational detail and more in the intellectual responsibility and human ingenuity underpinning the idea itself.

References

Arapoff, N. (1967). Writing: A thinking process. *TESOL Quarterly, 1*(2), 33–39.

Barclay, J. M. (2018). The letters of Paul and the construction of early Christian networks. *Letters and Communities: Studies in the Socio-Political Dimensions of Ancient Epistolography*, 289–301.

Bloom, H. (2014). *The western canon: The books and school of the ages.* Houghton Mifflin Harcourt.

Bukhari, S. W. R., Hassan, S. U., & Aleem, Y. (2024) Impact of artificial intelligence on copyright law: Challenges and prospects. *Journal of Law & Social Studies (JLSS)*, *5*(4), 647–656.

Cain, W. (2024). Prompting change: Exploring prompt engineering in large language model AI and its potential to transform education. *TechTrends*, *68*(1), 47–57.

Canbul Yaroğlu, A. (2024). The effects of artificial intelligence on organizational culture in the perspective of the hermeneutic cycle: The intersection of mental processes. *Systems Research and Behavioral Science*, *2024*, 1–13.

Castillo-Segura, P., Alario-Hoyos, C., Kloos, C. D., & Panadero, C. F. (2023, October). *Leveraging the potential of generative AI to accelerate systematic literature reviews: An example in the area of educational technology.* Paper presented at the 2023 World engineering education forum-global engineering deans council (WEEF-GEDC). IEEE.

Caulfield, J. (2023). *How to design and teach a hybrid course: Achieving student-centered learning through blended classroom, online and experiential activities.* Taylor & Francis.

Coeckelbergh, M., & Gunkel, D. J. (2024). ChatGPT: Deconstructing the debate and moving it forward. *AI & Society*, *39*(5), 2221–2231.

Cornish, W. (2010). The Statute of Anne 1709–10: Its historical setting. In K. C. L. Bently, H. Smith, U. Suthersanen, & P. Torremans (Eds.), *Global Copyright*. Edward Elgar Publishing.

Defoe, D. (2022). *The Cambridge edition of the Correspondence of Daniel Defoe.* Cambridge University Press.

De la Durantaye, K. (2007). The origins of the protection of literary authorship in ancient Rome. *BU International Law Journal*, *25*, 37.

Du Bois, J. W. (1987). Meaning without intention: Lessons from divination. *IPrA Papers in Pragmatics*, *1*(2), 80–122.

Duffy, B. K., & Winchell, M. R. (1989). "Speak the speech, I pray you." The practice and perils of literary and oratorical ghostwriting. *Southern Communication Journal*, *55*(1), 102–115.

Dunn, J. D. (2004). Book review: Paul, the law, and the covenant. *Interpretation, 58*(3), 319–321.

Eastman, D. L. (2021). The Pauline teachings of Peter in the Apocryphal Acts. *Neotestamentica, 55*(1), 89–109.

Farina, M., Lavazza, A., Sartori, G., & Pedrycz, W. (2024). Machine learning in human creativity: Status and perspectives. *AI & Society, 39,* 3017–3029.

Fathoni, A. F. C. A. (2023). Leveraging generative AI solutions in art and design education: Bridging sustainable creativity and fostering academic integrity for innovative society. *E3S Web of Conferences, 426,* 01102.

Ferris, D. R., & Hedgcock, J. S. (2023). *Teaching L2 composition: Purpose, process, and practice.* Routledge.

Foucault, M. (1969).What is an author? *Bulletin de la Société Française de Philosophie.*

Foucault, M. (2003). What is an author? In *Reading architectural history* (pp. 71–81). Routledge.

Freeman, M. A. (2023). *The hands that write: Life and training of Greco-Roman scribes* (Doctoral dissertation). Duke University.

Ganjavi, C., Eppler, M. B., Pekcan, A., Biedermann, B., Abreu, A., Collins, G. S., Gill, I. S., & Cacciamani, G. E. (2024). Publishers' and journals' instructions to authors on use of generative artificial intelligence in academic and scientific publishing: bibliometric analysis. *BMJ, 2024,* 384.

George, A. S. (2023). The potential of generative AI to reform graduate education. *Partners Universal International Research Journal, 2*(4), 36–50.

Giampieri, P. (2024). AI-powered contracts: A critical analysis. *International Journal for the Semiotics of Law-Revue internationale de Sémiotique juridique, 38*(2), 403–420.

Harper, K. (2012). Specters of Paul: Sexual difference in early Christian thought. *Journal of Late Antiquity, 5*(1), 216–218.

Hosseini, M., Rasmussen, L. M., & Resnik, D. B. (2023). Using AI to write scholarly publications. *Accountability in Research, 31*(7), 715–723.

Hutson, J., & Harper-Nichols, M. (2023). Generative AI and algorithmic art: Disrupting the framing of meaning and rethinking the subject-object dilemma. *Global Journal of Computer Science and Technology: D, 23*(1), 55–61.

Ilomäki, L., Lakkala, M., Kallunki, V., Mundy, D., Romero, M., Romeu, T., & Gouseti, A. (2023). Critical digital literacies at school level: A systematic review. *Review of Education, 11*(3), e3425.

Leung, T. I., de Azevedo Cardoso, T., Mavragani, A., & Eysenbach, G. (2023). Best practices for using AI tools as an author, peer reviewer, or editor. *Journal of Medical Internet Research, 25*, e51584.

Lyons, M. (2021). *Typewriter century: A cultural history of writing practices* (Vol. 46). University of Toronto Press.

Mazzi, F. (2024). Authorship in artificial intelligence-generated works: Exploring originality in text prompts and artificial intelligence outputs through philosophical foundations of copyright and collage protection. *The Journal of World Intellectual Property, 27*(3), 410–427.

Meeks, W. A. (2003). *The first urban Christians: The social world of the apostle Paul*. Yale University Press.

Merriam-Webster. (2024). Writing. *Merriam-Webster Dictionary.* https://www.merriam-webster.com/dictionary/writing

Mollick, Ethan. (2024, September 16). Scaling: The State of Play in AI: A brief intergenerational pause... *One Useful Thing.* https://www.oneusefulthing.org/p/scaling-the-state-of-play-in-ai

More, A. B. (2023). Implementing digital age experience marketing to make customer relations more sustainable. In Anand Nayyar, Mohd Naved, & Rudra Rameshwar (Eds.), *New Horizons for Industry 4.0 in modern business* (pp. 99–119). Springer International Publishing.

Moss, C. (2024). *God's ghostwriters: Enslaved Christians and the making of the Bible*. Little, Brown.

Myhill, D., Cremin, T., & Oliver, L. (2023). Writing as a craft: Re-considering teacher subject content knowledge for teaching writing. *Research Papers in Education, 38*(3), 403–425.

Nauta, M., Trienes, J., Pathak, S., Nguyen, E., Peters, M., Schmitt, Y., & Seifert, C. (2023). From anecdotal evidence to quantitative evaluation methods: A systematic review on evaluating explainable AI. *ACM Computing Surveys, 55*(13s), 1–42.

Nicholson, H. (2012). How to be engaging: Recreational reading and readers' advisory in the academic library. *Public Services Quarterly*, *8*(2), 178–186.

Noy, S., & Zhang, W. (2023). Experimental evidence on the productivity effects of generative artificial intelligence. *Science*, *381*, 187–192.

Perel, M., & Elkin-Koren, N. (2017). Black box tinkering: Beyond disclosure in algorithmic enforcement. *Florida Law Review*, *69*, 181.

Preiksaitis, C., & Rose, C. (2023). Opportunities, challenges, and future directions of generative artificial intelligence in medical education: scoping review. *JMIR Medical Education*, *9*, e48785.

Riley, L. A., & Brown, S. C. (1996). Crafting a public image: An empirical study of the ethics of ghostwriting. *Journal of Business Ethics*, *15*, 711–720.

Ritchhart, R., Church, M., & Morrison, K. (2011). *Making thinking visible: How to promote engagement, understanding, and independence for all learners.* John Wiley & Sons.

Roe, J., Renandya, W. A., & Jacobs, G. M. (2023). A review of AI-powered writing tools and their implications for academic integrity in the language classroom. *Journal of English and Applied Linguistics*, *2*(1), 3.

Rust, M. (2020). The role of the scribe: Genius of the book. In S. C. Akbari & J. Simpson (Eds.), *The Oxford handbook of Chaucer* (pp. 98–125). Oxford University Press.

Sarkar, A. (2023, June). Exploring perspectives on the impact of Artificial Intelligence on the creativity of knowledge work: Beyond mechanised plagiarism and stochastic parrots. Paper presented at the proceedings of the 2nd annual meeting of the symposium on human-computer interaction for work. ACM.

Schniedewind, W. M. (2024). *Who really wrote the Bible: The story of the scribes.* Princeton University Press.

Schoen, E. L. (1990). Anthropomorphic concepts of God. *Religious Studies*, *26*(1), 123–139.

Sinaga, P., & Feranie, S. (2017). Enhancing critical thinking skills and writing skills through the variation in non-traditional writing task. *International Journal of Instruction*, *10*(2), 69–84.

Stephens, E. (2023). The mechanical Turk: A short history of "artificial artificial intelligence." *Cultural Studies*, *37*(1), 65–87.

Stojanovic, L., Radojcic, V., Savic, S., Sandro, A., & Cvetkovic, D. S. (2023) The influence of Artificial Intelligence on creative writing: Exploring the synergy between AI and creative authorship. *International Journal of Engineering Inventions*, *12*(12), 70–74.

Techmeme Ride Home. (2025, January 9). Simon Willison and SWYX talk about the state of AI in 2025 [Video]. *YouTube*. https://www.youtube.com/watch?v=i4GIuFlDwiY

Thompson, M. P. (1993). Reception theory and the interpretation of historical meaning. *History and Theory*, *32*(3), 248–272.

Torppa, M., Niemi, P., Vasalampi, K., Lerkkanen, M. K., Tolvanen, A., & Poikkeus, A. M. (2020). Leisure reading (but not any kind) and reading comprehension support each other—A longitudinal study across grades 1 and 9. *Child Development*, *91*(3), 876–900.

Tsao, J., & Nogues, C. (2024). Beyond the author: Artificial intelligence, creative writing and intellectual emancipation. *Poetics*, *102*, 101865.

Turck, M. (2025, February 6). The AI coding agent revolution, the future of software, techno-optimism with Amjad Masad, CEO Replit. *The MAD Podcast*. https://www.youtube.com/watch?v=9xhDL2GbzaU

Virkler, H. A., & Ayayo, K. G. (2023). *Hermeneutics: Principles and processes of biblical interpretation*. Baker Books.

Wang, Y., Pan, Y., Yan, M., Su, Z., & Luan, T. H. (2023). A survey on ChatGPT: AI-generated contents, challenges, and solutions. *IEEE Open Journal of the Computer Society*, *4*, 280–302.

CHAPTER 4

Ethical and Practical Failures of Disclosure

When calculators first invaded classrooms in the 1970s, educators panicked that basic arithmetic skills would vanish; similarly, the arrival of spellcheck in Microsoft Word 95 sparked fears of literacy's collapse—yet math and spelling somehow survived. Today, academia faces comparable anxiety over generative artificial intelligence (AI) tools, responding with stringent disclosure mandates and detection tools like Turnitin's AI-hunting algorithms. Recent research claiming to identify linguistic fingerprints unique to each AI model—tracking quirks such as GPT-4o's frequent use of "utilize"—illustrates this ethical fixation. Yet this obsessive documentation misunderstands the fundamentally hybrid nature of creativity and reflects a culturally specific view of plagiarism sharply divergent from more communal global approaches. In our current, post-original landscape, bureaucratic insistence on disclosing every AI-generated contribution burdens creators unnecessarily, undermining rather than protecting true creativity.

4.1 Impracticalities and Ethical Misconceptions

The cat-and-mouse game between those using generative AI chatbots to produce undetectable text and those seeking to expose them has emerged as academia's latest flashpoint, fueling a perpetual conflict between students and faculty and between authors and publishers. Yet this fervor for detection and disclosure isn't novel. When spellcheck arrived in Microsoft Word 95, panic swept through educational circles, with educators prophesying the imminent decline of human spelling abilities. Similarly, in the mid-1970s, calculators prompted widespread anxiety that future adults would become incapable of performing basic arithmetic. In both cases, these tools were initially branded as aids for cheaters before becoming useful norms. The same story repeated with Grammarly in 2009, eliciting vehement objections from English professors and publishers wary of students—especially English as a second language(ESL) learners—and authors relying on algorithmic assistance. The current iteration of this ethical witch hunt is exemplified by anti-plagiarism software like Turnitin, recently repurposed as AI-detection tools, despite ironically beginning to integrate generative writing tools themselves. Recent research typifies the current mania, with studies like those by Stokel-Walker (2025) carefully cataloging the linguistic idiosyncrasies of large language models (LLMs)—GPT-4o's predilection for "utilize," Deep-Seek's frequent "certainly," and Claude's repetitive "according to the text." Yet the zeal to pinpoint generative authorship based on such minutiae fundamentally overlooks the messy reality of the creative process. This chapter thus explores the impracticalities and ethical misconceptions of obsessive disclosure require-ments, arguing that creativity—whether aided by spellcheck, calculators, or advanced language models—has always resisted bureaucratic attempts at strict documentation and control.

However, such acts of academic monitoring miss one major consideration: They do not identify the human intervention in the writing process and what was done by the writer during the process. This omission represents the most fundamental consideration — and the U.S. Copyright Office agrees—as the assumption typically made is that utilizing an LLM necessarily implies min-imal human input, perhaps nothing more sophisticated than a cursory prompt followed by a thoughtless act of copying and pasting. Equating this assump-tion to plagiarism is akin to dismissing an entire Renaissance poem due to the poet incorporating a line from Horace. In practice, even identical prompts yield distinctly varied outcomes across text, images, and video, undermining simplistic notions of replication. The algorithmic intricacies of each unique model and training set, characterized by probabilistic rather than deterministic outputs, inherently defy the possibility of exact duplication. Copyright law fundamentally seeks to protect originality—the unique aspects of a creative work—but contemporary generative models, by their very nature, obviate this possibility.

4.1.1 The Problem with Plagiarism

The complexities surrounding plagiarism within American academia and publishing predate the contemporary anxiety about generative writing. Histor-ically fraught with ambiguities, the definition and identification of plagiarism in the United States are marked by considerable debate and inconsistency. The *Oxford English Dictionary* defines plagiarism broadly as "the practice of copying another person's ideas, words or work and pretending that they are your own," highlighting actions ranging from overt copying to nuanced intellectual misappropriation. *Merriam-Webster* further underscores plagiarism's deceitful intent, characterizing it as "to steal and pass off ideas or words of another as one's own." Thus, whether executed overtly or through subtle manipulation, plagiarism is inherently an act involving conscious deception.

Within American institutions, numerous actions constitute plagiarism, from direct copying without attribution to subtler practices such as inadequate paraphrasing or summarizing without proper citation (Kumar et al., 2023; Putra et al., 2023; Taylor, 2023). Even minor alterations in sentence structure or the unauthorized use of various media sources—including audio, video, and expert interviews—can qualify as plagiarism. Additionally, practices like unauthorized reuse of one's own previously submitted work—commonly referred to as self-plagiarism—further complicate these boundaries (Kleebayoon & Wiwanitkit, 2023). Excessive collaboration or submitting identical essays for multiple courses without explicit approval also represent breaches of academic integrity. These expansive definitions underscore the complexity and contested nature of plagiarism within academic contexts.

The challenge then posed by generative technologies complicates plagiarism detection even further. Detection software routinely generates false positives, as noted, particularly disadvantaging neurodiverse individuals and nonnative English speakers. As Nadkarni (2024) reports, these often output a 100% AI use score, which is in itself a flag that it is unreliable, but also that neurodivergent students "think, learn, and process information differently from others." Therefore, the inherent variations and unpredictability in generative outputs highlight the impracticality of equating algorithmic assistance with outright intellectual theft. Plagiarism fundamentally aims to protect originality, yet, generative creation, by nature, defies precise reproduction due to probabilistic modeling and continual variation. Hence, copyright protections, premised upon uniqueness, are fundamentally misaligned with the operational realities of AI-generated content. Moreover, the insistence on exhaustive disclosure of generative processes represents not merely impractical bureaucratic excess but a profound misunderstanding of creativity itself. The granular documentation demanded by institutions like the U.S. Copyright Office fails to recognize the inherently iterative, subconscious, and spontaneous nature of creative endeavors.

However, concepts of academic integrity and the avoidance of plagiarism are far from universal. The American interpretation, which defines plagiarism fundamentally as passing off another's ideas as one's own, diverges significantly from perceptions in other global cultures, where ethical implications and definitions can vary considerably. In East Asian contexts such as China, Japan, and South Korea, the communal approach to knowledge ownership often diminishes the necessity for individual attribution (West, 2006). This perspective arises from deeply rooted collectivist cultures and Confucian teachings, wherein memorizing and reiterating authoritative texts signify respect and acknowledgment rather than intellectual appropriation. Students in these societies are socialized into viewing knowledge as a collective asset, potentially clashing with Western conventions of individual authorship and formal citation (Grudecki, 2021). Similarly, within Middle Eastern contexts, informal quoting and paraphrasing of revered religious or political texts without explicit citation is commonplace, operating under the assumption that these materials constitute universally recognized common knowledge (Lapidus, 2022). Formal citation in such cases could inadvertently imply ignorance or disrespect, negatively impacting students' academic perceptions and standings.

Eastern European nations, although maintaining formal citation frameworks, generally adopt a more lenient stance toward plagiarism, placing less emphasis on rigid adherence to standards of academic integrity (Makarova, 2019). Comparable attitudes emerge in Latin American educational systems, where academic misconduct, including plagiarism, often lacks the rigorous enforcement and explicit discourse prevalent within the United States (Ison, 2018). Conversely, Western European nations, along with Canada and Australia, align closely with U.S. standards, emphasizing individual responsibility and adherence to stringent anti-plagiarism guidelines (Leask, 2006; Stoesz & Eaton, 2022). These varying cultural perspectives have historically presented substantial challenges for international students and scholars within the American academia (Amsberry, 2009; Gunnarsson et al., 2014). The emphasis on individual intellectual property rights and strict citation practices prevalent in the United States frequently generates confusion and frustration among students and scholars originating from cultures where such conventions are either absent or distinctly conceptualized. The integration of generative technologies amplifies these complexities, introducing novel dimensions to the creation, attribution, and dissemination of scholarly content. Given this evolving scenario, academia and publishers must continually reassess and refine their definitions and approaches to plagiarism.

However, considerations of scholarly integrity and plagiarism not only differ across cultures but also vary significantly within a single culture, as professional norms and expectations diverge widely among different academic and professional fields. Within this context of the relationship between education and scholarship in one's trained field, it is critical to examine the evolving landscape of professional ethics. Traditionally, educational objectives closely mirrored workforce demands, dictating not only curricular content but also shaping scholarly research and the values underpinning academic disciplines. However, the rapid technological transformation—marked by automation, algorithmic decision-making, and shifting workplace roles—necessitates reconsideration of what is taught, valued, and assessed within educational institutions (Hutson et al., 2024). The increasingly agential and uncertain world renders existing educational frameworks increasingly obsolete, demanding a radical rethinking of academic and professional ethics.

4.1.2 Ethics as Historically Contingent/Created

At the same time, what is deemed "ethical" within scholarly and artistic creation has undergone profound transformations. During the Renaissance, for instance, overt quotation of ancient philosophers, especially Aristotle referred to as "The Philosopher," without explicit citation was considered perfectly ethical, even admirable (Grafton, 1985). Authors displayed their erudition through seamless integration of classical wisdom, inviting readers to delight in recognizing intertextual references, thereby validating their intellectual sophistication through (Plett, 1999). Similarly, after Michelangelo's monumental achievements, it became ethically acceptable—indeed expected—to incorporate visual and thematic allusions to his seminal works like the Sistine Chapel frescoes, the Medici Chapel, or sculptures such as Moses and David. Court artists like Bronzino (1503–1572) would take this to its extreme in a "Where's Waldo" like fashion in the *Martyrdom of Saint Lawrence* (1569) (Figure 4.1) where numerous figures are all taken from Michelangelosque precedents with Adam from the Sistine Chapel front and center. This form of "interpictoriality," akin to Shakespearean intertextuality, functioned less as plagiarism and more as cultural homage, reinforcing shared intellectual heritage and visual literacy among audiences.

The very notion that artistic theft, originality, and the ethics surrounding copying are recent cultural constructs finds compelling validation in the Baroque discourse

Figure 4.1: Bronzino, *Martyrdom of Saint Lawrence*, 1569, fresco. San Lorenzo, Florence (CC O)

of repetition. As Loh (2004) notes, originality as an immanent criterion for judging artistic value is itself an invention of the eighteenth century; thus, to retroactively impose this framework upon earlier periods is anachronistic at best. Prior to this shift, repetition—rather than being denigrated as derivative—was celebrated as demonstrative imitation, a strategic practice intended to invoke recognition and to foreground intertextual dialogues between artists and informed viewers. The Italian Baroque period, especially through artists such as Alessandro Varotari

(Padovanino) (1588–1649), explicitly embraced such repetition to generate new meanings and aesthetics from established sources, thereby highlighting the paradox that "originality" was frequently rooted in imitative acts.

In fact, for Baroque theorists and practitioners, the ethical valuation of "theft" in art hinged not on whether one appropriated visual or thematic material, but rather on how effectively — and ingeniously — this appropriation was executed. Artistic judgment focused on the viewer's ability to recognize subtle references and appreciate the artist's mastery in transforming previous compositions. Marco Boschini (1602–1681), a prominent seventeenth-century Venetian critic, celebrated Padovanino precisely for his adept pastiches, praising them as ingenious compositions that intentionally alluded to Titian's (ca.1488/90-1576) Bacchanals and other canonical works. Boschini referred admiringly to Padovanino as the "Vice-Author," acknowledging and even honoring the painter's skill in crafting sophisticated visual mosaics from historical references (Loh, 2004, p. 482).

What contemporary critics might condemn as "theft" was often extolled by Baroque theorists as *acutezza*—an intellectual wit that delighted informed audiences by embedding layered meanings and visual metaphors within artworks. Emanuele Tesauro, a leading theorist of this rhetorical strategy, emphasized metaphor as a method of simultaneously expressing and concealing multiple meanings. Wit, in Tesauro's formulation, lay in the viewer's pleasurable discovery of hidden references and the joy derived from decoding the subtle layers of repetition embedded in artistic works (Loh, 2004). However, while pastiche and repetition initially enjoyed a privileged position as valid artistic strategies, the eighteenth-century reevaluation of originality led to a transformation of taste and ethical judgment. The once-celebrated practice of artistic repetition gradually shifted toward negative associations, culminating in terms like pasticcio carrying connotations of tasteless imitation or stylistic indigestion. Critics such as the French philosopher and writer Denis Diderot (1713–1784) lamented this emergent trend, observing that pastiche had become a mark of contempt — ironically dissuading artists from engaging with the traditions that had long sustained artistic creativity.

Around the same time in the Romantic period (1780–1850) one continued to see shifts in this ethical and professional paradigm both in literary and artistic works, introducing the modern notion of "originality," closely linked to the concept of the solitary, often male, "genius." This Romantic ideal elevated unique creative output above all else, casting derivative works in an ethically dubious light (Fredriksson, 2007). Yet contemporary scholarship and artistic practice have effectively dismantled this Romantic myth, acknowledging that all creative endeavors inherently build upon prior knowledge and cultural context (Mitter, 2008). Postmodernism, in particular, embraces pastiche and conceptual appropriation — further developing from artists like Andy Warhol (1928–1987), who openly recontextualized familiar imagery, and literary figures such as Thomas Pynchon (1937–) and Toni Morrison (1931–), who utilized extensive references and appropriations as integral components of their narratives. These approaches highlight how the ethics of originality have evolved, recognizing the value in recontextualizing existing elements rather than generating wholly unprecedented ones (Fokkema, 2024).

4.1.3 The Educational Paradox of Documentation and Creativity

Despite this, contemporary education for artists and scholars places significant emphasis on meticulously documenting creative and scholarly processes — an approach both burdensome and strikingly novel. Students today find themselves asked to log reflections, maintain detailed portfolios, and submit exhaustive records of their iterative journeys. Practices such as reflective journaling and process portfolios purportedly nurture transparency and critical engagement, though one might quip they seem designed more to appease assessors than inspire genuine creativity (Lam, 2024). Ironically, this trend toward exhaustive documentation coincides with a widespread acknowledgment that absolute originality is an elusive, if not mythical, goal. Indeed, educational institutions reinforce an inherent paradox: demanding rigorous records of processes that openly acknowledge their derivative nature. Moreover, integrating AI into fields like medicine has markedly redefined ethical standards; in contemporary healthcare, neglecting to consult AI diagnostic tools is practically malpractice (Kalra et al., 2024). The implications are clear — ignoring algorithmic guidance in critical decision-making scenarios has become ethically dubious. Thus, as technological advancements and shifting cultural norms continually reshape professional ethics, academia must remain agile. Core concepts of originality, authorship, and ethical scholarly conduct must undergo constant reassessment to retain relevance amid rapid global transformations. Understanding this evolution is essential for navigating the complex landscape of academic integrity today and preparing students for an ethically complex future.

Yet the latest disruption in education is hardly unprecedented — technological shifts have long reshaped educational values and methodologies, prompting constant reinvention (often reluctantly) of what counts as ethical knowledge production. Indeed, the educational landscape has frequently shifted in response to emerging societal needs, with each technological leap heralding a new set of professional expectations. Consider the humble codex, innovated during the late Roman Empire (fourth–fifth centuries CE): Its adoption streamlined scholarly communication, effectively birthing the educational structures suited to administrative and governance roles (Harnett, 2017). Not to be outdone, the fifteenth-century printing press democratized information access, transforming widespread illiteracy into an intellectual faux pas during the Renaissance (and consequently, making ignorance far less excusable at dinner parties). Subsequently, the nineteenth-century chalkboard standardized pedagogical practices, shaping the uniform skill sets necessary for the disciplined labor of the Industrial Revolution (Li, 2023). Fast-forward to the mid-twentieth century, when calculators swiftly moved classrooms beyond the drudgery of manual arithmetic toward complex analytical thinking, better aligned with the technical complexity demanded by postindustrial economies (Monaghan et al., 2016). Each epoch's advancements not only remedied contemporary educational deficiencies but fundamentally redefined ethical parameters for knowledge creation and dissemination. Hence, ethical standards are not static; they are historically contingent and continuously remade in the crucible of technological and cultural change—an insight that compels contemporary educators and professionals to approach generative technologies not with trepidation, but with informed adaptability.

Building on this historical trajectory, it is essential to consider early learning systems, which relied heavily on oral traditions for the preservation and transmission of knowledge, fundamentally shaping educational practices and societal values within preliterate cultures (Gardner, 2003; Llewellyn & Ng-A-Fook, 2017). The eventual adoption of written text—enabled significantly by innovations such as the codex—marked a profound shift away from memory dependence, fostering literacy skills essential for long-term preservation and analytical engagement with information (Leu, 1982). This transition empowered learners to independently access and critically analyze information, yet it simultaneously exposed deep disparities in access to literacy, compelling education systems to confront emergent inequalities. Later technological breakthroughs continued reshaping educational priorities in alignment with workforce demands: The printing press radically democratized access to knowledge, fostering self-directed learning—a trajectory significantly extended by contemporary digital technologies, emphasizing digital literacy and adaptive problem-solving skills (Susilo et al., 2023). Nevertheless, persistent challenges like inequitable access and diminished human-mediated instruction highlight the ongoing need for education to carefully balance technological adoption against critical engagement, alongside evolving understandings of ethical reasoning (Moravec & Martínez-Bravo, 2023). This historical interplay underscores education's dual function: preparing individuals economically while simultaneously advancing broader societal progress.

Ancient Greek education further illustrates this foundational evolution, laying significant groundwork for contemporary pedagogical models—most notably through the Socratic method. The term "pedagogy" itself derives from the Greek *paidagogos,* originally denoting a slave entrusted with escorting children to educational institutions and overseeing their instruction. This historical context reveals deep-seated foundations within Greek society, where education transcended mere intellectual advancement, serving as a vehicle for instilling civic virtues and moral accountability (Djurayevich, 2021). The Socratic method, formulated by Socrates (470/469 BCE–399 BCE), relied profoundly upon dialogue, mentorship, and disciplined memory as core educational tools. Through rigorous questioning and structured reflective inquiry, this method addressed the pressing societal imperative to develop critical thinkers and future civic leaders, embodying intellectual rigor combined with ethical consciousness (Benson, 2011).

Rather than rote memorization typical of earlier educational paradigms, the Socratic method positioned dialogue and mentorship at its core, emphasizing analytical reasoning and reflective engagement. Socrates famously exemplified this through probing dialogues, notably his extensive exploration of justice in Plato's Republic (375 BCE). By systematically challenging participants' assumptions and leading them toward deeper self-awareness and intellectual independence, Socrates reshaped the educational exchange into a collaborative, dynamic interaction rather than a unidirectional transmission of knowledge (Ogilvy, 1971). A vivid demonstration of this pedagogical shift appears in Socrates' interaction with Euthyphro, in which structured questioning dismantled superficial beliefs about piety, driving learners toward logical consistency and critical reflection (Furley, 1985). Such active, discourse-driven engagement set crucial precedents for modern educational practices, emphasizing active learning and intellectual exploration (Lee, 2014). Ultimately, ancient Greek education reveals how ethical considerations around knowledge transmission were historically contingent—placing less emphasis on proprietary claims and more on moral virtue, rhetorical authenticity, and communal intellectual responsibility, thus profoundly influencing the evolving ethical landscape of pedagogy and scholarship.

While elements of the Socratic method remain influential in contemporary educational practices, the method itself—like other pedagogical innovations—is fundamentally tethered to the specific societal values and workforce imperatives of its historical moment. Ancient Greek society placed a premium on critical thinking and ethical deliberation, qualities indispensable for governance and civic participation; thus, Socratic dialogue served a distinctly practical purpose, training future statesmen in the nuanced art of political leadership. The method's emphasis on mentorship further reinforced hierarchical structures: After all, who better to instruct the emerging elite than their predecessors, steeped in the intricacies of power? But here's the paradox: Despite Socrates' famed interrogations promoting intellectual rigor, the method itself remained distinctly exclusive, catering predominantly to the privileged few (Reeve, 2003). As with most advancements, access hinged on one's place within a stratified social order, rendering many segments of the population mere observers—left outside the circle of inquiry and insight. This exclusionary dynamic underscores a persistent tension in educational history: Pedagogical innovations frequently mirror the prejudices and hierarchies inherent to the societies producing them, rather ironically limiting the transformative potential they purportedly offer. Even as oral traditions gave way to written texts, ostensibly democratizing knowledge, similar patterns of marginalization persisted, revealing how technological progress and social inequities frequently collude to restrict rather than expand educational access. From this historical vantage, the creative ethics underlying educational practices in patriarchal societies become strikingly clear: Innovation is rarely neutral. Instead, it reflects, reinforces, and, at times, amplifies existing inequalities, crafting intellectual ecosystems as much defined by who they exclude as by whom they elevate. Thus, the Socratic method, celebrated for championing reason, simultaneously illuminates an ethical contradiction at the heart of creativity itself—one must always ask, who precisely does this creativity serve?

4.1.4 Orality to Literacy: How Technology Dictates Learning

Regardless, the shift from oral tradition to written text heralded a significant leap forward in knowledge transmission, with the codex serving as an undeniable technological catalyst. While oral traditions thrived on memory, communal consensus, and the ever-shifting artistry of storytellers, writing froze knowledge in time, rendering it portable, durable, and—perhaps terrifyingly for traditionalists—immutable (Jovchelovitch, 2019). The codex (Figure 4.2) represented a rather elegant evolution: Individual pages, crafted meticulously

Figure 4.2: Codex Egberti (980-993), fol. 13: *Nativity of Jesus, the Annunciation to the shepherds.* Schatzkammer of Stadtbibliothek Trier (Germany) (CC 4.0)

from papyrus, parchment, or vellum, were securely bound together, superseding the clumsier scroll format prevalent in earlier eras. Developed during the waning days of the Roman Empire around the fourth century CE, this innovation quickly gained prominence due to its superior capacity for storage, accessibility, and ease of use (Harnett, 2017). No longer confined to linear unrolling, readers now flipped freely through pages, quickly locating sections with efficiency previously unimaginable—thus began humanity's enduring romance with the physical book. Educational systems and institutions seized upon this capability, codifying religious doctrines, bureaucratic edicts, and scholarly treatises in structured, permanent forms, thereby reshaping both teaching and learning. Knowledge dissemination expanded exponentially, transcending immediate communities and time-bound oral performances to penetrate broader societal frameworks. Yet this technological advancement also reshaped the very nature of authorship itself. Writing in codex form distanced authors from audiences, conferring upon texts an independent existence divorced from immediate interpersonal contexts. No longer mere repositories of communal wisdom, authors emerged as individuated authorities, their voices preserved, scrutinized, and revered long after their physical selves had vanished. The codex did more than merely preserve and transmit knowledge; it altered profoundly the creative act itself, assigning to the author a new cultural weight, prestige, and accountability, marking the transition from ephemeral storyteller to enduring intellectual figure.

Historical figures such as Saint Augustine vividly illustrate the transformative influence wielded by the codex—indeed, Augustine's *Confessions* (397–400 CE) serve as a veritable case study of how reading shifted from communal performance to a deeply private and contemplative pursuit (O'Donnell, 2012). No longer was knowledge merely enacted aloud, ephemeral and open to collective scrutiny; instead, texts became fixed, reliable companions facilitating sustained introspection, individual interpretation, and silent dialogue between the author and the reader. This newfound privacy invited deeper intellectual engagements—precisely the type of inward meditation Augustine famously cherished—reshaping reading into an intensely personal act of self-discovery. Simultaneously, the codex fostered formal, structured educational frameworks, setting the stage for defined curricula, standardized practices, and the meticulous codification of specialized knowledge (Brown, 2020). As texts became tangible, replicable, and organized, knowledge ceased to be fluidly communal and evolved into carefully delineated domains, complete with authoritative gatekeepers who controlled interpretation, legitimacy, and access. In doing so, the codex gave rise to institutionalized learning environments, libraries, and scholarly networks: veritable fortresses of expertise wherein specific figures—authors, theologians, and philosophers—assumed unprecedented authority, their voices amplified and enshrined through the permanence of written form. Such specialized learning undoubtedly reinforced authorial hegemony, granting select individuals an authoritative monopoly over fields of knowledge, interpretation, and ideological frameworks. While ethically ambiguous, this shift inevitably concentrated intellectual power among a scholarly elite who defined orthodoxy, censured dissent, and determined the parameters of acceptable thought. Yet, in a rather ironic twist, the ethical quandary was twofold: Though undeniably exclusionary, authorial dominance also safeguarded intellectual rigor, authenticity, and fidelity to sources—preventing dilution and distortion of ideas. The ethical landscape of specialized learning under codex hegemony was intricately paradoxical, raising the perennial scholarly question: Does institutional control of knowledge serve primarily to protect intellectual integrity, or to entrench structures of power and privilege?

The transformation from oral to written culture, epitomized by the codex's widespread adoption, resonates profoundly with the scholarly insights of Walter J. Ong and Marshall McLuhan—both of whom dissected how technological advancements in communication reshape cognition, societal structures, and, notably, ethical paradigms surrounding knowledge creation. Walter J. Ong, in his influential text *Orality and Literacy: The Technologizing of the Word* (1982), posits that writing externalized memory, enabling analytical abstraction and detached reasoning previously inconceivable in purely oral cultures. However, Ong does not merely celebrate literacy's liberating potential; instead, he carefully acknowledges the ethical complexities this transition entailed. Writing imposed an ethical shift: Whereas oral societies favored communal validation and collaborative construction of truth, literacy introduced an ethos of individual intellectual ownership, emphasizing accuracy, originality, and private accountability. In oral traditions, knowledge was inherently collective, shared, and ethically evaluated by community consensus. Writing and the codex altered this communal orientation, institutionalizing individual authorship and implicitly prioritizing solitary mastery of ideas.

Similarly, Marshall McLuhan, in *The Gutenberg Galaxy* (1962), identifies writing—and later print—as a transformative "technology of the intellect," revolutionizing human communication and dramatically reshaping cultural values, including ethical frameworks surrounding authorship and information dissemination. According to McLuhan, the codex's linearity and repeatability established new intellectual hierarchies, bestowing upon authors unprecedented power to define canonical truths and control interpretive authority. This shift carried profound ethical implications: Authors emerged as privileged gatekeepers, ethically responsible for both accuracy and the social impact of their works, and simultaneously empowered and constrained by their textual permanence. Knowledge was no longer fluid, subject to dynamic reinterpretation, but etched indelibly upon pages, reifying particular viewpoints and, perhaps troublingly, marginalizing alternatives. McLuhan underscores this ethical paradox: The codex amplified individual accountability and integrity of authorship yet simultaneously solidified intellectual hierarchies, shaping knowledge production to reflect—and often reinforce—existing power structures. Thus, both Ong and McLuhan compellingly illustrate that while writing and the codex dramatically expanded cognitive possibilities, this technological evolution simultaneously reframed the ethics of knowledge creation, privileging certain voices, marginalizing others, and irrevocably altering humanity's collective intellectual trajectory.

These technological and pedagogical shifts reshaped the interplay between education and work, solidifying literacy as the essential foundation for occupational preparation within increasingly specialized economies. Writing facilitated the standardization of curricula, transforming education into a systematic enterprise capable of training scholars, bureaucrats, and professionals whose expertise underpinned emerging societal institutions. Consider medieval monastic schools: Codices became indispensable instruments, meticulously employed to impart theology, philosophy, and canon law — crucial knowledge enabling clergy to navigate the labyrinthine demands of spiritual guidance and ecclesiastical administration. Herein, as Walter J. Ong and Marshall McLuhan persuasively argue, literacy did not merely enhance memorization; it external-ized memory, empowering abstract reasoning and critical analysis — precisely the cognitive tools demanded by professions reliant upon managing complex, systematic knowledge (McLuhan, 1962; Ong, 1982). Such intellectual capabilities directly translated into occupational competencies, from interpreting intricate theological texts to performing detailed administrative duties. Indeed, research underscores how formal literacy instruction, anchored in written texts, directly bolstered skills such as advanced reading comprehension and systematic data management — aptitudes that rapidly became indispensable within administra-tive, scholarly, and commercial fields (Mikulecky, 1982). Yet this codification also introduced ethical dilemmas, institutionalizing hierarchies by creating a specialized elite class whose privileged access to knowledge perpetuated societal inequalities. Thus, education's increasingly systematic alignment with workforce demands simultaneously democratized opportunities and reinforced established power structures, compelling societies to continually reexamine ethical boundaries around inclusion and access.

The widespread adoption of the chalkboard in the nineteenth century repre-sented a critical turning point, reshaping educational practices amid profound societal transformations driven by industrialization. Invented by James Pillans around 1801 and popularized at institutions such as West Point Academy by 1809, the chalkboard enabled standardized, visual instruction for entire classrooms, transitioning education from individualized, rote memorization to collaborative learning environments (George & Pandey, 2024; Jones, 2024). This transition fundamentally altered classroom dynamics, allowing teachers to simultaneously present complex ideas through diagrams and illustrations, enhancing student engagement and comprehension — particularly in mathematics and science — by enabling real-time interaction (Vadakedath et al., 2018). Crucially, the chalk-board supported emerging societal values of inclusivity and equitable access to education, embodying the ethos that education constituted a public good rather than an elite privilege (Collins, 1971; Justman & Gradstein, 1999). However, this technological advancement also raised ethical questions by aligning education closely with industrial imperatives: The chalkboard facilitated the cultivation of a literate, numerate workforce proficient in standardized tasks, emphasizing conformity and efficiency over individualized intellectual development (Stevens, 1995). Moreover, the shift toward publicly visible student work amplified anxieties among students and resistance among teachers accustomed to personalized instruction methods, underscoring how pedagogical innovations simultaneously expanded and constrained educational experiences (Vadakedath et al., 2018). Concerns over chalk dust also arose, highlighting the unintended consequences of widespread technology adoption, though affordability and adaptability ultimately secured its place as a cornerstone educational tool (Sekar et al., 2021).

The industrialization of education, exemplified by innovations such as the chalkboard, simultaneously undermined romantic notions of the solitary, specialized creator. Previously, mastery of specialized knowledge had provided certain societal segments with distinct economic advantages through exclusive apprenticeships and private mentorship. Industrialized education, by contrast, democratized access to foundational skills through standardized, visual group instruction, diminishing individual knowledge monopolies and reshaping ethical considerations around intellectual ownership and access. This transformation significantly impacted academic culture, laying the groundwork for contemporary expectations in research and publication. Emerging predominantly in the nineteenth-century German university system, standards such as rigorous methodological rigor, peer-review processes, specialization, and the prioritization of original scholarship became entrenched (Becker et al., 2011). These criteria persist today as benchmarks defining academic legitimacy and institutional prestige, reinforcing specialized expertise even as educational technologies evolve toward inclusivity and accessibility.

Each technological innovation — from codices and chalkboards to contemporary digital tools — redefines educational priorities and reflects broader societal demands, altering the skills deemed essential by successive generations (Khan, 2024). The codex, initially supporting literacy and individual contemplation, later gave way to collaborative visual learning through chalkboards, projectors, and ultimately digital whiteboards. More recently, computers and Internet technologies shifted educational focus toward media literacy, independent learning, and critical evaluation skills essential in a digitally interconnected world (Grimalt-Álvaro et al., 2019; Reguera & López, 2021). Despite these innovations, adoption remains uneven, shaped by institutional resources, cultural attitudes, and persistent educational inequities, reinforcing the complex ethical dynamics between technology, knowledge democratization, and societal power structures (Winzenried et al., 2010).

4.1.5 Necessary Disruption: AI and Institutions

In the first century CE, apostles dictating their messages to scribes likely did not envision sparking ethical debates about algorithmic authorship nearly two millennia later—but here we are. Just as those ancient scribes subtly shaped sacred texts through interpretive nuance, generative platforms now mediate human intention through layers of computational abstraction, embedding biases from vast, training datasets (Teixeira da Silva & Tsigaris, 2023). Humans have rarely operated in intellectual isolation, despite romantic notions to the contrary; authorship has long been collaborative, iterative, and ethically complicated. Yet today's calls for stringent disclosure of generative technology use in scholarly or artistic outputs feel historically unprecedented—indeed, somewhat absurd—given authorship's inherently collaborative legacy. Demanding such disclosures ironically ignores the reality of cognitive interdependence between human and machine collaborators, reducing nuanced creation to mere algorithmic citation.

The philosophical tension underlying these debates—humanistic ideals versus vocational pragmatism—has dogged educational discourse for centuries, echoing from classical liberal arts curricula through contemporary science, technology, engineering, and mathematics (STEM)–focused imperatives (Chan, 2016). Advocates of traditional, humanistic approaches often resist generative tools, concerned about diluting originality or diminishing ethical rigor. Conversely, utilitarian proponents pragmatically embrace these technologies, provided students still demonstrate measurable skill mastery (Abulibdeh et al., 2024). Publishers now reflect this divide explicitly in their author guidelines: humanities journals require transparency regarding generative use—or outright forbid it—while STEM-oriented publishers pragmatically endorse integration, provided rigorous standards remain intact. Thus, the pendulum swings once again between romantic notions of individual human creativity and utilitarian acknowledgment of augmented cognition.

Nevertheless, generative AI is unquestionably reshaping education and industry, democratizing and complicating knowledge creation in equal measure. Tools like ChatGPT allow students to swiftly synthesize summaries, craft persuasive essays, or interpret data, prompting educators to shift from emphasizing mere information retrieval toward fostering critical skills in algorithmic literacy (Khan, 2024; McCarthy & Yan, 2024). Of course, while these technologies simplify processes, they create novel ethical dilemmas: Students must now actively identify and critique algorithmic biases and inaccuracies, a requirement unimaginable to students relying solely on traditional texts or face-to-face lectures (Chiu et al., 2024).

In parallel, industries such as journalism and media production increasingly rely on AI to automate tasks formerly executed by human professionals, reshaping job markets and prompting uncomfortable ethical reflections (Aleessawi & Alzubi, 2024; Smith & Johnson, 2024).

Nowhere is this ethical friction more glaring than in fields historically defined by human expertise. Healthcare management increasingly delegates tasks — from patient communications to content creation — to intelligent platforms, raising ethical considerations about human oversight and employment security (Syed & ES, 2023). Software developers similarly find themselves displaced, as generative algorithms rapidly generate reliable code, challenging traditional team structures and reshaping careers overnight (Marquis et al., 2024). And insurance customer service — once a bastion of empathetic human interaction — is now efficiently managed by chatbots, leaving thousands pondering the morality of technology-enabled redundancy (Khalisa, 2024). These realities underscore that human cognitive limitations in memory, processing speed, and accuracy inevitably position the latest generation of models as an ethically necessary companion rather than an optional luxury.

At each technological pivot point — from the codex to digital whiteboards and now generative AI — educational priorities inevitably recalibrate, reflecting evolving societal and economic needs (Grimalt-Álvaro et al., 2019; Khan, 2024; Reguera & López, 2021). Pedagogies have repeatedly adapted, shifting focus between independent literacy, collaborative learning, technical proficiency, and, now, nuanced algorithmic interpretation. Each shift presents ethical challenges regarding inclusivity, access, and the distribution of intellectual authority — questions made especially urgent today as generative tools democratize, yet complicate, creative production and scholarship. In fact, authorship has rarely ever reflected pure individualism; from apostles collaborating with scribes to monks in scriptoria (Figure 4.3) copying ancient texts to scholars relying on libraries full of secondary literature, the myth of solitary intellectual labor has long been exaggerated. Autonomous systems further dissolve that myth, foregrounding the inherently collaborative nature of authorship: Humans guide algorithms, and algorithms assist humans, in a continual feedback loop. Nonetheless, the Copyright Office and publishers' insistence on explicit disclosure of technological contributions often seems to deny this reality, ironically positioning technology as an external intruder rather than an integral participant in contemporary creativity.

Such demands for transparency risk misunderstanding the fundamental shift toward co-intelligence that generative platforms represent. Physicians already

Figure 4.3: Jean Le Tavernier, *Scribe*, after 1456 (CC O)

openly acknowledge ethical obligations to employ AI precisely because of documented accuracy advantages compared to unassisted human judgment (McCarthy & Yan, 2024). Such ethical necessity extends seamlessly into research and writing: Human scholars simply cannot independently review, synthesize, and interpret the sheer volume of relevant sources that new models effortlessly manage. Thus, generative tools move from being ethically questionable shortcuts to indispensable collaborators—acknowledged openly or not—rendering stringent disclosure requirements somewhat impractical, if not outright impossible.

This transformation compels educators to realistically prepare students for actual practices they will encounter professionally. Rather than clinging to idealized notions of purely human-driven scholarship, curricula must intentionally develop algorithmic literacy, ethical discernment, and interpretative agility. Students need training to critically analyze automated outputs, discerning embedded biases, inaccuracies, and ethical implications (Chiu et al., 2024; Hutson & Ceballos, 2023). Moreover, education should cultivate uniquely human strengths—emotional intelligence, cultural sensitivity, and creativity—abilities

AI struggles to replicate but are essential in mediating technological outputs thoughtfully and responsibly (Shaikh, 2025).

In short, the future of education and creativity lies not in resisting generative platforms but embracing their nuanced capabilities, limitations, and ethical complexities. Teaching students (future authors and artists) to interact productively with AI—to interpret algorithmic outputs critically and responsibly—equips them for genuine professional environments. The scholar or artist of tomorrow will not be a solitary creator but an adept collaborator, navigating and harnessing the productive tensions between human ingenuity and artificial efficiency. Such individuals, prepared through integrated, realistic curricula, promise to lead meaningfully within the new, dynamic ecosystems of human and artificial intelligence.

4.2 The Bureaucratic Burden of Documenting Authorship

In the halcyon days of scholarly and creative production, the author's chief task was deceptively straightforward: create something remarkable, then claim the credit (and perhaps survive the inevitable scholarly criticism). Today, however, creators find themselves largely participating in writing as administrative performance, forced to document each micro-step of their creative process—often ironically at the expense of the creativity these measures purport to safeguard. Modern intellectual property policies, with their well-meaning but burdensome insistence on transparency, push creators toward an uncomfortable bureaucratic ritual. This ritual increasingly shifts attention from substantive intellectual engagement to an obsessive accounting of one's intellectual process as though the scholarly act itself depended more on bureaucratic bookkeeping than intellectual rigor.

This section explores the hidden consequences of this documentation imperative, examining how the demand for meticulous transparency reshapes the very act of creation itself. What began as a reasonable request to clarify authorship in the era of algorithmic collaboration now risks reducing authors to meticulous archivists rather than imaginative thinkers. Transparency, designed to illuminate, paradoxically threatens to bury intellectual labor under layers of detailed minutiae, creating more confusion than clarity. Thus, creators increasingly spend precious intellectual bandwidth documenting, categorizing, and certifying their every creative step—a burdensome and at times darkly comic endeavor, akin to authors becoming their own bureaucratic biographers. Consequently, while the pursuit of transparency may reassure institutional anxieties about originality, it inadvertently encourages a compliance culture, in which creative expression becomes subordinate to procedural fidelity. As we unpack the performative rituals, ironic obscurities, and psychological toll of this bureaucratic hypervigilance, we will ultimately consider pragmatic alternatives to protect methodological privacy—without sacrificing accountability or, indeed, our collective sanity.

4.2.1 The Performative Nature of Documentation Rituals

Mandatory authorship documentation has increasingly become a kind of academic Kabuki—a stylized performance that assuages institutional fears more than it enriches scholarship. Researchers diligently fill out contribution forms, check compliance boxes, and draft reflexive statements, knowing fully well that much of this theater is about *appearing* accountable. The process is often more per-formative than informative. For instance, in clinical settings a safety checklist intended to improve outcomes was observed to turn into a *ceremonial event* giving only an "illusion of compliance" while actual quality control suffered (Safety & Performance Research Summaries [SPRS], 2024). In other words, the act of documenting can become a proxy for real integrity, a comfort blanket for institutions. Universities and publishers, anxious about ethics and oversight, mandate exhaustive authorship statements and data management plans. Scholars oblige by going through the motions—sometimes sincerely, sometimes with a roll of the eyes—because failing to do so would raise red flags. The net effect is a growing disconnect between intent and outcome. Like actors playing roles, authors learn the script of bureaucratic correctness (often with a touch of cyn-icism), ensuring every required field is filled and every signature is in place. It satisfies the auditors and administrators who can then proclaim due diligence. But these rituals rarely spark new insights or enhance the truthfulness of the work. Instead, they risk devolving into *hollow formalities*. The integrity that documentation is supposed to uphold is undermined by the very *ritualism* of the practice—a classic case of process over purpose. In sum, documentation rituals today often function as appeasement performances, calming institutional nerves through sheer volume of paperwork. Everyone involved can say "proper procedures were followed," even if nothing of substance has changed in the underlying research. It's a satirical twist: The academy, devoted to knowledge, sometimes values looking *procedurally proper* over actually being insightful. And most scholars, with a knowing sigh, play along in this paper pageant, hoping the show of compliance will keep the real bureaucratic wolves at bay.

4.2.2 The Bureaucratic Irony: Transparency as Obscurity

In an ideal world, transparency shines a light on the research process, but excessive detail in documentation can create a blizzard of information that obscures the very insights it was meant to illuminate. The pursuit of total transparency can lead to opacity. A key example comes from clinical recordkeeping: Doctors, empowered by digital forms, began recording every detail, producing bloated medical files. The result? Critical facts get lost in voluminous notes. As Chaiken dryly noted for the Health Data Management Report (2016), "excessive documentation leads to bloated records that obscures important information." In research, we see a similar pattern. Ethics protocols and methodology appendices have swelled in length and complexity over the years. In the realm of human-subjects research, for example, informed consent forms that once were a page or two have ballooned into booklets. Over the last few decades, the average consent document grew by a factor of ten, often reaching twenty to forty pages—a length *proven* to impair participant understanding (Wisgalla & Hasford, 2022). The requirement for exhaustive disclosure (meant to ensure participants are fully informed) instead leaves them overwhelmed and under-informed, unable to identify the truly vital points from the legalese and technical minutiae. This is the bureaucratic irony at work: In trying to leave nothing unsaid, we say *so much that nothing is heard*. A similar phenomenon haunts scholarly writing. Method sections and data availability statements sometimes drown readers in irrelevant procedural trivia, obscuring the elegant simplicity of a study's core idea. When every keystroke and data transformation is logged, the narrative of discovery can disappear under sediments of detail. The *spirit* of transparency—clarity and honesty—is undermined by an *overdose* of transparency. In effect, the signal-to-noise ratio worsens. Both practitioners and readers face what one might dub *transparency fatigue*, unable to see the forest of insight through the trees of documentation. The solution, ironically, may require *stepping back* on disclosure: Knowing what not to document is as important as knowing what to lay bare. Otherwise, our attempts at openness risk plunging us into a new form of obscurity—one *written in fine print* and appendices.

4.2.3 Shifting Accountability: From Creator to Documentarian

One of the strangest unintended consequences of today's bureaucratic demands is how they subtly recast the role of the author. Scholars used to be valued primarily as creators—thinkers, experimenters, and writers. Now they must also be hyper-efficient documentarians, accountable for an endless trail of paperwork about their work. The center of gravity of accountability shifts from *what* you produced to *how well you logged every step producing it*. Consider the modern researcher's daily grind: updating grant portals, entering metadata into institutional repositories, filling progress reports, compliance checklists, Institutional review board (IRB) modifications, data management plan templates—a litany of administrative upkeep. It's not just anecdotal whining; the numbers bear it out. A recent analysis in the United Kingdom reported by Jones (2022) estimated that researchers and staff waste around 55,000 person-days per year on redundant data entry—retyping the same information about publications, grants, and projects into various systems. That's 150 years of human effort annually, merely pushing paper (or pixels) around! Little wonder many academics quip that *"actual research"* has become a side hobby. As one commentary lamented, scholars often feel they can only do real research in their "spare time," outside of normal working hours consumed by admin duties. In effect, the creative role is squeezed into the margins as the bureaucratic role expands. Accountability has been reframed: A "good" researcher is one who never misses a form deadline, whose every dataset has a DOI, and whose lab notebook is impeccably time-stamped—even if those measures collectively nibble away at the time and mental energy available for creative thought. The tragedy is that content creation and documentation are not given equal weight. Miss a paperwork step, and institutional alarms go off; miss a bold idea, and likely no compliance officer will notice. This shift in priorities incentivizes scholars to allocate their efforts accordingly. We are witnessing the rise of the scholar–administrator hybrid, an author who is as much a clerk as a creator. The paper trail has become as important as the paper itself. While accountability is crucial, in overcorrecting for past laxities the system risks turning brilliant minds into glorified recordkeepers. It's an academic echo of the joke: *"I love being a writer, except for all the paperwork."* Except now, it's no joke—it's the daily reality for many researchers navigating the modern compliance maze.

4.2.4 The Rise of Compliance Culture in Creative Scholarship

Out of these trends emerges a compliance culture that permeates creative scholarship. Rules and guidelines multiply, and with them comes a mindset that values *playing it safe* over pushing the envelope. When every step must be justified, recorded, and submitted for approval, scholars naturally begin to self-censor and self-regulate to avoid rocking the boat. The ethos shifts: from bold inquiry to cautious compliance. We see this in how research agendas are set. If an unconventional project promises great insights but doesn't fit neatly into the predefined compliance checkboxes, a risk-averse scholar (or their department) might think twice about pursuing it. After all, grant proposals and ethics applications reward those who check all the required boxes; there's little room on those forms for wild ideas or methodological improvisation. Over time, this conditions a conservative approach to inquiry. Studies that meet established norms sail through easier, whereas edgy or cross-disciplinary work faces extra scrutiny (and the bureaucratic gauntlet that comes with it). Evidence of this culture appears in academic systems around the world. The so-called *audit culture* in universities that Chow (2015) noted is to "keep academics compliant, obedient, and domesticated," leaving "little breathing room" for research that falls outside narrow, easily measurable benchmarks. In such an environment, activities that are hard to quantify — say, community-engaged scholarship, exploratory blue-sky experiments, or interdisciplinary art–science fusions — tend to be discouraged or devalued. One scholar observed that academics are being "driven away" from socially engaged or innovative pursuits precisely because those endeavors "can't be easily captured in the audit forms" used to evaluate performance (Chow, 2015). The result is a perverse incentive: do work that *scores well on metrics* and bureaucratic evaluations, even if that's not the work that truly advances knowledge. The compliance-oriented mindset also diminishes intellectual risk-taking at the individual level. A young researcher, keenly aware of the labyrinth of rules, may opt for a safe dissertation topic that won't tangle them in red tape or controversy. A seasoned professor might shelve a creative but unorthodox project, dreading the paperwork nightmare it could entail. Over time, the habit of coloring within the lines becomes second nature. Scholars internalize the compliance officer that looks over their shoulder. The irony is rich: Universities — historically hotbeds of radical thought and paradigm shifts — now often nurture a culture akin to a regulatory agency, where everyone is trying to avoid making a mistake. Compliance culture, of course, has its rationales (preventing fraud, ensuring ethics, etc.), but when it crowds out the spirit of adventure, it comes at a steep cost: the slow stifling of creativity under the weight of its own safeguards.

4.2.5 Authorial Identity Reduced to Data Points

Who is an author in the age of hyper-documentation? Increasingly, the answer comes back not as a rich narrative of ideas and stylistic voice, but as a profile of data points. In the drive to catalog every aspect of authorship, we risk flattening the nuanced identity of creators into a series of metrics, identifiers, and checklist items. Today's author might be known by an ORCID iD number, a h-index, a grant income total, and a list of bureaucratic compliances ("100% of required training completed!") more readily than by the actual content of their work. In evaluative settings, this reductionist view is stark. Hiring committees, award panels, and ranking systems often rely on quantifiable proxies for quality: number of publications, impact factors of journals, citation counts, and so on. The rich tapestry of a scholar's contributions—mentoring students, public outreach, creative thought experiments, even the *style* and *originality* of their prose—can fade out of the picture. Audit-driven academia has distort[ed] scholars' work by tabulating academic worth through the simplest algorithm, focusing almost solely on easily countable outputs like refereed publications, citation indices, and grant dollars. Whole realms of intellectual endeavor that resist quantification (think monographs in the humanities, experimental art pieces, or long-term foundational research that bears fruit only decades later) get marginalized, because they don't neatly convert into immediate data points. Even the process of authorship itself is diced into checkbox categories. Consider the now-common "author contribution" statements in journals: Author A did X%, Author B did Y%, each role narrowly defined (conceptualization, data curation, etc.). These are meant to give credit where it's due—a noble goal—but in practice they can feel like reducing a complex collaboration into a pie chart of duties. The dynamic interplay of coauthors (where ideas bounce around, merge, and mutate) cannot really be captured by a few standardized labels. Yet the bureaucracy demands *formalized attribution*. The texture of *how* an insight emerged or *why* an author's perspective is unique gets lost; identity is boiled down to a set of functions performed. Furthermore, an author's *intent* and *creative vision*—those subtle, often unspoken drivers of why a work took the shape it did—have no entry field in a compliance form. A surrealist poet's quirky worldview, a scientist's hunch based on years of tacit knowledge, an engineer's imaginative leap—these are not data points, and, thus, they often disappear from the official record. Instead, authorship gets checkpointed: Did you submit your disclosure form? Check. Did you follow the reporting guideline? Check. Congratulations, you are an *officially recognized author*. It's a bit like reducing a master chef's identity to a food safety certificate and the calorie counts of their dishes. All necessary information, perhaps, but hopelessly inadequate in conveying the person behind the work. In sum, the bureaucratic lens tends to see authors as collections of attributes and outputs, not as creative intellects with depth and idiosyncrasy. This flattening of identity serves certain institutional needs (evaluation, attribution, and accountability, foremost), but it impoverishes the narrative of scholarship. It leaves us with authors as data and metadata—easier to catalog, perhaps, but far less inspiring to celebrate.

4.2.6 Algorithmic Bureaucracy: Automating Transparency

Faced with the untenable burden of manual documentation, academia has turned its hopeful eyes toward algorithms and automation. If humans are drowning in paperwork, why not let the machines take some of the load? Thus emerges the era of algorithmic bureaucracy, where software and AI tools are enlisted to automatically log, check, and even generate documentation. On paper (or rather, on screen), this promises salvation: less drudgery for scholars, more consistent recordkeeping, and perhaps even real-time compliance monitoring that flags issues before they become problems. *What could go wrong?* In some ways, these tools are indeed alleviating burdens. Consider the integration of research infor-mation systems using persistent identifiers. No longer does a researcher need to enter the same publication data into five different portals; with systems talking to each other via ORCID or DOI metadata, one update propagates everywhere. The U.K. review reported on by Jones (2022) of research bureaucracy concluded that widespread use of such PIDs is "a key part of the solution" to redundant admin work, allowing funder, university, and publisher systems to interoperate. In practical terms, that means fewer mind-numbing hours uploading PDFs or pasting references. Likewise, automation is making inroads in other fields with similar issues. In medicine, for example, AI-driven documentation assistants are being used to transcribe and organize notes so doctors can focus on patients. A recent implementation of a voice-to-text AI tool in Brazil aimed to "*reduce the bureaucratic burden on doctors*" by automatically generating structured medical records from a conversation, de Paula et al. (2025) report. By analogy, one can imagine AI tools that record laboratory workflows or annotate code changes for a researcher—a kind of automated lab notebook that spares the scientist from constant manual note-taking. These developments are promising, but they in-troduce new layers of complexity and ethical ambiguity. For one, automation is only as good as the rules we give it. If the underlying bureaucratic requirements are flawed, automating them can just entrench the flaws faster. There's also the risk of overreliance: Will scholars become disengaged from the documentation process entirely, losing situational awareness of important details because "the system handles it"? An automatically generated record might be complete but mindless — it takes human insight to note which decision was trivial and which was a turning point. Moreover, algorithmic tools can create a false sense of security. Just as a spellchecker doesn't ensure good writing, an automated transparency tool doesn't ensure *understanding*. We might get beautifully version-controlled research logs that no one ever reads or learns from. There's also the question of new burdens: maintaining the tools themselves. Anyone who has wrestled with a poorly designed institutional software (or suffered an automated compliance system's bugs) knows that "labor-saving" technology can sometimes create its own headaches. And, of course, ethical issues lurk — especially with AI — regarding privacy (e.g., auto-tracking every keystroke of a researcher could be seen as a surveillance overreach) and accountability (if an AI logs something incorrectly, who is responsible?). In short, algorithmic bureaucracy is a double-edged sword. It can *streamline* bureaucracy, but it can also formalize it to an unprecedented degree. The key will be to ensure these tools are used to *serve* scholarly cre-ativity rather than to further shackle it. We must be wary of ending up with "automated micromanagement" — a scenario where researchers are hemmed in by machine-enforced protocols. Used wisely, automation could free up time for actual research; used poorly, it could tighten the bureaucratic web. The hope is that we get more of the former: less time on paperwork, more on ideas. But as any IT admin will tell you with a sigh, simply adding tech doesn't guarantee simplification — sometimes it just relocates the complexity.

4.2.7 The Psychological Cost of Bureaucratic Hypervigilance

Perhaps the most heartbreaking cost of all this is the psychic toll on the creators themselves. The joy of scholarship — that feeling of being "in the zone" when insights flow — is increasingly at odds with a state of constant bureaucratic hy-pervigilance. How can one *freely chase ideas* when a little voice (often wearing an administrator's hat) is perpetually whispering, "Did you document that? Did you get approval for this? Don't forget to log that change!" The cognitive load of incessant self-monitoring can turn the creative process into a stilted stop–start grind. Instead of diving into research with abandon, scholars tiptoe through a minefield of requirements, afraid that a single oversight could come back to haunt them. This atmosphere breeds anxiety — the sense that one must be on guard at all times. Psychologists might liken it to an elevated stress baseline: never fully relaxed, the mind always partially occupied with what form or rule is lurking around the corner. Over time, this can erode the passion that drives creative work, leading to exhaustion and burnout. The connection between bureaucratic burden and burnout is well documented (no pun intended). In medicine, Swanson (2021) noted that today "doctors spend more time on electronic record keeping than with patients," and unsurprisingly this *bureaucratic burden is implicated in professional burnout*. Academia shows parallel symptoms. Professors report feeling like glorified administrators rather than researchers or teachers, fueling a quiet crisis of morale. Morris' (2020) candid essay in *University Affairs* pointed out how top-down administrative overload "breeds confusion, frustration, anger and resentment" among faculty. These are strong words—*frustration, anger*, and *resentment*—but they resonate with many who feel their creative spark dimming under the cold fluorescent light of compliance culture. The psychological cost also manifests in more subtle ways. Hypervigilance in documenting every decision can rob researchers of spontaneity. Imagine a jazz musician who, after every riff, had to stop and fill out a form explaining which notes they just played and why. The music would die, replaced by mechanical justification. Similarly, if a scientist must cross every *t* and dot every *i* in real time, serendipitous exploration—the kind that leads to unexpected breakthroughs—is less likely to happen. The brain simply won't wander far from the well-trodden, rule-safe path. Young academics, especially, report a kind of paralysis at times: the fear of doing something wrong bureaucratically that leads them to second-guess experimental approaches or avoid unconventional ideas. This psychological weight is not always easy to see from the outside. It creeps in gradually, a cumulative effect of years of navigating compliance. But its impact is real: a diminution of intellectual bravery and a creeping sense of scholarly drudgery. When creativity starts to feel like filling in blanks on forms, something has gone deeply wrong. The human factor—enthusiasm, curiosity, and the willingness to take risks and even to fail gloriously—is what propels knowledge forward. We must reckon with the fact that bureaucracy, when overbearing, saps this very human spirit. The cost is measured in the research not pursued, the ideas not born, and the innovations unrealized—all because the people who carry the torch of knowledge are too busy carrying the clipboard of compliance, or too tired to lift the torch in the first place.

4.2.8 Toward a Pragmatic Framework for Methodological Privacy

Is there a way to have our cake and eat it too—to uphold accountability without strangling creativity? The answer may lie in embracing a more pragmatic framework that explicitly preserves *methodological privacy* and intellectual autonomy, even as it maintains legitimate transparency. In practice, this means carving out spaces of *freedom* within the research process, where scholars can experiment, make mistakes, and think aloud *off the record*, so to speak. At the same time, it means being smarter and more selective about what truly needs documenting and reporting. Below are some guiding principles that could help balance these aims:

- *Purpose-Driven Documentation*: Every piece of documentation required should have a clear rationale that serves the *common good* of scholarship (integrity, reproducibility, and ethics). If a form or report doesn't demonstrably improve the research's reliability or social responsibility, think hard about whether it's needed. As one set of best-practice guidelines puts it, *"All policies, regulations, rules and forms should improve the common good, and clearly document how they do so"* (Morris, 2020). In short, document with purpose, not for its own sake.

- *Minimally Intrusive Oversight*: Design accountability checks that do not detract from a researcher's ability to perform their fundamental duties. This could mean, for example, requiring a concise methods summary and data deposit *after* a project is complete, rather than mandatory exhaustive logs *during* the creative process. Allow researchers to keep rough work, preliminary analyses, or creative brainstorming private. Oversight should focus on endpoints (is the published result credible and ethical?) more than policing every interim step. By keeping the spotlight on outcomes and critical decisions, we respect the "flow" of research and writing, intervening only where it truly counts.

- *Protected Exploratory Space*: Encourage institutions to grant researchers a *protected sandbox* for idea development. Not every hypothesis scribbled on a whiteboard needs to be archived; not every off-the-cuff test requires official clearance. This is not to advocate anarchy or lack of accountability, but to recognize, as Michael Polanyi famously did, that *"we can know more than we can tell."* Creative work involves tacit intuition and trial-and-error that often cannot be fully articulated in formal documentation. A pragmatic framework would acknowledge the limits of codification—some aspects of discovery are best left to the tacit dimension, at least until they crystallize into something worth reporting. Methodological privacy means giving researchers the *trust and space* to develop ideas without immediate external scrutiny, much like how writers might draft multiple chapters before showing anyone, or how artists keep sketchbooks for their eyes only.

- *Streamlined and Nonredundant Processes*: Bureaucracy should serve the scholar, not the other way around. Where documentation is truly necessary, it must be as efficient as possible. Eliminate duplicate requests for information (no researcher should have to enter the same data twice across different forms—modern database systems can ensure that). If a new compliance requirement is added, see if an old one can be removed or combined, to avoid endless accretion. Keep forms short and jargon-free whenever feasible. In essence, make the administrative parts of research feel less like torture. The easier and clearer the process, the less it will weigh on the mind and time of the researcher—and the more likely it is to actually be done well rather than done resentfully.

- *Support over Surveillance*: Shift the tone of oversight from a policing model to a support model. Instead of inundating scholars with "gotcha" checkpoints, provide resources to help comply with sensible standards. This could mean dedicated administrative assistants or automated tools (calibrated to be user-friendly) to handle much of the clerical load, freeing the researcher to focus on content. It also means training and guidance that frame documentation as *part of the research ecosystem*—not a separate bureaucratic gauntlet. When possible, let specialists handle specialized tasks, for example, data curators assisting with data management plans, or librarians helping with citation compliance. "Only bureaucrats and administrators should do the work of bureaucrats and administrators," as one rule Morris (2020) puts forth states bluntly. In practice, that means leveraging professional staff and infrastructure to shoulder administrative tasks, rather than foisting everything on the scholars themselves.

- *Proportional Accountability*: Not every project carries the same risks or stakes, so a one-size-fits-all documentation is overkill. A pragmatic framework would calibrate requirements to the level of complexity and potential impact. A short theoretical paper might not need a 10-page methods appendix; a simple classroom study shouldn't require the same review rigor as a clinical trial. Develop tiers of review and documentation so that accountability is right-sized. This keeps smaller projects nimble and doesn't discourage innovative pilot studies with red tape, while still ensuring that big, high-impact, or sensitive projects get the thorough oversight they warrant.

- *Regular Sunset and Review*: Bureaucratic rules have a way of sticking around "just because." It's healthy to periodically review documentation requirements and cull the obsolete ones. Introduce sunset clauses for new policies—they expire unless actively renewed based on demonstrated need. Solicit feedback from those in the trenches: If a particular form is universally despised and yields no clear benefit, be willing to toss it. In other words, keep the bureaucracy *lean and accountable* to its users. By pruning the thicket, you maintain only what truly adds value, preventing the buildup of "paper chalk" that clogs the wheels over time.

In putting these principles into action, continuous dialogue between scholars and institutional administrators becomes essential. The conversation, however, should avoid devolving into a bureaucratic staring contest. Instead, the aim must be mutual clarity: Accountability frameworks exist to empower scholarship, not suffocate it under endless procedural formalities. When both sides appreciate that basic truth, a reasonable equilibrium emerges. Researchers accept measured oversight as a necessary component of academia's social contract; administrators, in turn, acknowledge that dissecting every creative impulse under a regulatory microscope achieves little beyond filling filing cabinets and fostering frustration.

Indeed, this pragmatic approach neatly supports the authors' larger argument against exhaustive disclosure in the deployment of AI-driven tools for content creation. Just as micromanaged bureaucratic processes drain scholarly energy—turning creative minds into reluctant archivists—so too do excessive demands for transparency in AI use risk suffocating the very innovation these tools promise to nurture. The "trust but verify" mantra, therefore, should be proportionate rather than obsessive: Trust scholars to exercise intellectual autonomy when engaging generative technologies, verify selectively and sensibly, and remember always that the ultimate goal remains discovery and innovation, not impeccably organized footnotes or algorithmic audit trails.

This balanced perspective doesn't promise to eradicate all bureaucratic friction—some red tape stubbornly clings to academic life like gum to a shoe—but it can certainly keep administrative procedures from overshadowing scholarly substance. The re-centering of trust, proportional oversight, and intellectual flexibility allows academia to safeguard its most precious resource: the human imagination and the freedom to harness it. Without such freedom, even the most meticulous documentation yields nothing but neatly ordered stagnation. Let us strive to ensure our scholarly ecosystems remember this foundational truth, enabling authors once again to create first, document sensibly afterward, and avoid reducing the vibrant art of scholarship to an elaborate game of administrative box-ticking.

Therefore, this chapter has critically examined the complex relationship between institutional demands for exhaustive documentation of authorship and the realities of intellectual labor, particularly in an age dominated by generative AI. The analysis presented makes clear a paradox central to contemporary scholarship: that relentless pursuit of transparency, intended to clarify creative processes, frequently yields the exact opposite outcome. Instead of illuminating authorship, excessively detailed recordkeeping tends to obscure meaningful human contributions beneath a labyrinthine archive of trivial documentation. Like a map so intricately drawn it becomes unreadable, hyper-documentation risks reducing clarity to opacity, and insight to bureaucratic noise.

Further compounding this issue is the performative dimension inherent in modern documentation rituals. As demonstrated, these practices often function less as genuine guarantors of intellectual integrity and more as institutional displays designed to assuage anxiety over accountability. Under the veneer of meticulous oversight, academic creativity risks transformation into an administrative performance—a dance choreographed not by intellectual curiosity but, instead, by bureaucratic mandate. The chapter's exploration underscores how such performative routines effectively shift the scholar's role from creator to curator, substituting genuine intellectual innovation with meticulous record management. Ironically, this transition risks eroding the very scholarly qualities—risk-taking, originality, and exploration—that academic institutions profess to champion. Additionally, this discussion highlighted how rigorous compliance cultures, driven by institutional anxieties about technological advancements, flatten nuanced human identities and creative efforts into simplistic data points. Authorial complexity, historically celebrated for its interpretive richness, risks reduction to checkboxes and performance metrics—barely distinguishable from the algorithmic mechanisms institutions purport to regulate. Paradoxically, automated tools designed to simplify documentation can further entrench these bureaucratic complexities, introducing new ethical ambiguities even as they claim to resolve administrative burdens.

This chapter has addressed a less tangible yet profoundly consequential issue: the psychological toll exacted by bureaucratic hypervigilance. The demand to record every incremental decision not only disrupts cognitive flow but also fosters an atmosphere of anxiety antithetical to creativity itself. As illustrated, scholars constrained by the expectation of continuous surveillance may hesitate to pursue unconventional lines of inquiry, ultimately curbing the spontaneous intellectual leaps vital to groundbreaking work. In order to mitigate these deleterious effects, we advocate for a balanced and pragmatic approach to methodological privacy — one grounded in proportionality and reason. While acknowledging the necessity of legitimate oversight, the chapter proposes recalibrating expectations toward sensible accountability. By adopting a policy framework centered on a measured principle — "trust, but verify within reason" — institutions can preserve creative freedom, intellectual autonomy, and human dignity. Documentation, then, becomes a targeted act, clearly purposeful rather than habitually ritualistic.

In sum, this chapter underscores a critical message: The bureaucratic impulse, left unchecked, threatens to eclipse intellectual creativity entirely. In attempting to ensure transparency, accountability, and authenticity, institutions must remain ever mindful of the true endgame — innovation, discovery, and meaningful scholarly contribution. Only through deliberate recalibration can academia ensure that documentation practices serve their intended supportive role, safeguarding rather than suppressing the most precious scholarly resource of all: the human imagination, free to explore and innovate unencumbered by excessive administrative burdens.

References

Abulibdeh, A., Zaidan, E., & Abulibdeh, R. (2024). Navigating the confluence of artificial intelligence and education for sustainable development in the era of industry 4.0: Challenges, opportunities, and ethical dimensions. *Journal of Cleaner Production, 437*, 140527.

Aleessawi, N. A. K., & Alzubi, S. F. (2024). The implications of Artificial Intelligence (AI) on the quality of media content. *Studies in Media and Communication, 12*(4), 41–51.

Amsberry, D. (2009). Deconstructing plagiarism: International students and textual borrowing practices. *The Reference Librarian, 51*(1), 31–44.

Becker, S. O., Hornung, E., & Woessmann, L. (2011). Education and catch-up in the industrial revolution. *American Economic Journal: Macroeconomics, 3*(3), 92–126.

Benson, H. H. (2011). Socratic method. In D. Morrison (Ed.), *The Cambridge companion to Socrates* (pp. 179–200).

Brown, M. P. (2020). Codex. In D. S. Lynch (Ed.), *Oxford research encyclopedia of literature*. Oxford University Press.

Chaiken, B. (2016, December 15). Why it's time to reconsider a documentation vision for EHRs. Health Data Management. https://www.healthdatamanagement. com/articles/why-its-time-to-reconsider-a-documentation-vision-for-ehrs

Chan, R. Y. (2016). Understanding the purpose of higher education: An analysis of the economic and social benefits for completing a college degree. *Journal of Education Policy, Planning and Administration, 6*(5), 1–40.

Chiu, T. K., Ahmad, Z., Ismailov, M., & Sanusi, I. T. (2024). What are artificial intelligence literacy and competency? A comprehensive framework to support them. *Computers and Education Open, 6*, 100171.

Chow, S. (2015, September 1). Higher education's silent killer. *Riarpatch.* https:// briarpatchmagazine.com/articles/view/higher-educations-silent-killer#:~:text=Meant%20to%20keep%20academics%20compliant%2C,and%20Indigenous%20 forms%20of%20knowing

Collins, R. (1971). Functional and conflict theories of educational stratification. *American Sociological Review, 36*, 1002–1019.

de Paula, P. A. B., Severino, J. V. B., Berger, M. N., Veiga, M. H., Ribeiro, K. D. P., Loures, F. S., Todeschini, S. A., Roeder, E. A., & Marques, G. L. (2025). Improving documentation quality and patient interaction with AI: A tool for transforming medical records—An experience report. *Journal of Medical Artificial Intelligence, 8*, 1–16.

Djurayevich, A. J. (2021). Education and pedagogy. *Journal of Pedagogical Inventions and Practices, 3*, 179–180.

Fokkema, A. (2024). *Postmodern characters: A study of characterization in British and American postmodern fiction* (Vol. 4). Brill.

Fredriksson, M. (2007). The Avant-Gardist, the male genius and the proprietor. *Nordlit, 11*(1), 275–284.

Furley, W. D. (1985). The figure of Euthyphro in Plato's "dialogue." *Phronesis, 30*(2), 201–208.

Gardner, P. (2003). Oral history in education: Teacher's memory and teachers' history. *History of Education, 32*(2), 175–188.

George, A. S., & Pandey, D. (2024). The evolution of education as a tool for corporate utility: From industrial revolution to present-day vocational preparation. *Partners Universal International Innovation Journal, 2*(4), 01–12.

Grafton, A. (1985). Renaissance readers and ancient texts: Comments on some commentaries. *Renaissance Quarterly, 38*(4), 615–649.

Grimalt-Álvaro, C., Ametller, J., & Pintó, R. (2019). Factors shaping the uptake of ICT in science classrooms: A study of a large-scale introduction of interactive whiteboards and computers. *International Journal of Innovation in Science and Mathematics Education, 27*(1), 18–36.

Grudecki, M. R. (2021). Plagiarism as a culturally-motivated crime. *Asian Journal of Law and Economics, 12*(3), 237–252.

Gunnarsson, J., Kulesza, W. J., & Pettersson, A. (2014). Teaching international students how to avoid plagiarism: Librarians and faculty in collaboration. *The Journal of Academic Librarianship, 40*(3–4), 413–417.

Harnett, B. (2017). The diffusion of the codex. *Classical Antiquity, 36*(2), 183–235.

Hutson, J., & Ceballos, J. (2023). Rethinking education in the age of AI: The importance of developing durable skills in the Industry 4.0. *Journal of Information Economics, 1*(2), 9.

Hutson, J., McMaken, W. T., & Vosevich, K. (2024). From codex to code: Pedagogical transformations in the age of technological innovation. *International Journal of Recent Engineering Science, 11*(6), 198–215.

Ison, D. C. (2018). An empirical analysis of differences in plagiarism among world cultures. *Journal of Higher Education Policy and Management, 40*(4), 291–304.

Jones, P. (2022, December 5). Unnecessary research bureaucracy is killing academic productivity, but it IS fixable. *The Scholarly Kitchen.* https://scholarly-kitchen.sspnet.org/2022/12/05/unnecessary-research-bureaucracy-is-killing-academic-productivity-but-it-is-fixable/#:~:text=research%20that%20we%20at%20MoreBrains,it%20has%20real%20consequences%20for

Jones, S. (Ed.). (2024). The mechanics' institutes and the spread of "useful knowledge." In *Manchester minds: A university history of ideas* (pp. 23–38). Manchester University Press.

Jovchelovitch, S. (2019). *Knowledge in context: Representations, community and culture.* Routledge.

Justman, M., & Gradstein, M. (1999). The industrial revolution, political transition, and the subsequent decline in inequality in 19th-century Britain. *Explorations in Economic History, 36*(2), 109–127.

Kalra, N., Verma, P., & Verma, S. (2024). Advancements in AI based healthcare techniques with FOCUS ON diagnostic techniques. *Computers in Biology and Medicine, 179*, 108917.

Khalisa, A. (2024). The digitalization in insurance broker industry: How artificial intelligence affect this industry. *Ilomata International Journal of Management, 5*(1), 261–279.

Khan, M. Z. (2024). AI revolutionizing content diversity and cultural sensitivity in India. *International Journal of Cultural Studies and Social Sciences, 2024*, 124–130.

Kleebayoon, A., & Wiwanitkit, V. (2023). Artificial intelligence, chatbots, plagiarism and basic honesty: Comment. *Cellular and Molecular Bioengineering, 16*(2), 173–174.

Kumar, V., Verma, A., & Aggarwal, S. P. (2023). Reviewing academic integrity: Assessing the influence of corrective measures on adverse attitudes and plagiaristic behavior. *Journal of Academic Ethics, 21*(3), 497–518.

Lam, R. (2024). Understanding the usefulness of e-portfolios: Linking artefacts, reflection, and validation. *International Review of Applied Linguistics in Language Teaching, 62*(2), 405–428.

Lapidus, I. M. (Ed.). (2022). *Middle Eastern cities: A symposium on ancient, Islamic, and contemporary Middle Eastern urbanism.* University of California Press.

Leask, B. (2006). Plagiarism, cultural diversity and metaphor — Implications for academic staff development. *Assessment & Evaluation in Higher Education, 31*(2), 183–199.

Lee, C. P. (2014). *The effects of interactive discourse, the Socratic method, and active learning labs on student achievement at the university level–a comparative approach* [Doctoral dissertation].

Leu, D. (1982). Differences between oral and written discourse and the acquisition of reading proficiency. *Journal of Literacy Research, 14*, 111–125. https://doi.org/10.1080/10862968209547440

Li, L. (2023). Revisiting "blackboard": Transformation of medium, space and pedagogy in school education. *Educational Philosophy and Theory, 55*(7), 773–786.

Llewellyn, K. R., & Ng-A-Fook, N. (Eds.). (2017). *Oral history and education: Theories, dilemmas, and practices*. Springer.

Loh, M. H. (2004). New and improved: Repetition as originality in Italian baroque practice and theory. *The Art Bulletin, 86*(3), 477–504.

Makarova, M. (2019). Factors of academic misconduct in a cross-cultural perspective and the role of integrity systems. *Journal of Academic Ethics, 17*(1), 51–71.

Marquis, Y. A., Oladoyinbo, T. O., Olabanji, S. O., Olaniyi, O. O., & Ajayi, S. A. (2024). Proliferation of AI tools: A multifaceted evaluation of user perceptions and emerging trend. *Asian Journal of Advanced Research and Reports, 18*(1), 30–35.

McCarthy, K. S., & Yan, E. F. (2024). Reading comprehension and constructive learning: Policy considerations in the age of Artificial Intelligence. *Policy Insights from the Behavioral and Brain Sciences, 11*(1), 19–26.

McLuhan, M. (1962). *The Gutenberg galaxy: The making of typographic man*. University of Toronto Press.

Mikulecky, L. J. (1982). Job literacy: The relationship between school preparation and workplace actuality. *Reading Research Quarterly, 17*(4), 400–422.

Mitter, P. (2008). Decentering modernism: Art history and avant-garde art from the periphery. *The Art Bulletin, 90*(4), 531–548.

Monaghan, J., Trouche, L., Borwein, J. M. (Eds.). (2016). The calculator debate. In *Tools and Mathematics* (pp. 305–331).

Moravec, J. W., & Martínez-Bravo, M. C. (2023). Global trends in disruptive technological change: Social and policy implications for education. *On the Horizon: The International Journal of Learning Futures, 31*(3/4), 147–173.

Morris, D. (2020, July 17). Death by download: The problem of misguided bureaucracy. *University Affairs*. https://universityaffairs.ca/opinion/death-by-download-the-problem-of-misguided-bureaucracy/#:~:text=The%20failure%20to%20consult%20reveals,significantly%20from%20those%20of%20faculty

Nadkarni, A. (2024, October 25). Neurodivergent students more likely to be flagged by AI detectors. *AIDP*. https://blog.aidetector.pro/neurodivergent-students-falsely-flagged-at-higher-rates/

O'Donnell, J. J. (2012). *Augustine confessions: Augustine confessions: Volume 1: Introduction and text* (Vol. 1). Oxford University Press.

Ogilvy, J. A. (1971). Socratic method, Platonic method, and authority. *Educational Theory, 21*(1), 3–16.

Ong, W. J. (1982). *Orality and literacy: The technologizing of the word.* Methuen.

Plett, H. F. (1999). Rhetoric and intertextuality. *Rhetorica, 17*(3), 313–329.

Putra, I. E., Jazilah, N. I., Adishesa, M. S., Al Uyun, D., & Wiratraman, H. P. (2023). Denying the accusation of plagiarism: Power relations at play in dictating plagiarism as academic misconduct. *Higher Education, 85*(5), 979–997.

Reeve, C. D. C. (2003). The Socratic movement. In R. Curren (Ed.)*, A companion to the philosophy of education* (pp. 7–24). Blackwell Publishing.

Reguera, E. A., & López, M. (2021). Using a digital whiteboard for student engagement in distance education. *Computers & Electrical Engineering, 93,* 107268.

Safety & Performance Research Summaries. (2024, February 10). Surgical checklists behaving badly…new study suggests they can result in ritualistic practices decoupled from their core goals. *Wiley.* https://safety177496371.wordpress.com/2024/02/10/surgical-checklists-behaving-badly-new-study-suggests-they-can-result-in-ritualistic-practices-decoupled-from-their-core-goals/

Sekar, A., Varghese, G. K., & Varma, R. M. K. (2021). Occupational exposure to particulate matter during blackboard teaching and its deposition in the airways of human lungs. *International Archives of Occupational and Environmental Health, 94*(8), 1963–1974.

Shaikh, Z. P. (2025). Artificial intelligence-based emotional intelligence and effective leadership: Applications, implications, and ethical bias. In Z. Sayed (Ed.), *Emotionally intelligent methods for meaningful leadership* (pp. 223–254). IGI Global Scientific Publishing.

Smith, D., & Johnson, S. (2024). AI-powered automation: Impacts on workforce dynamics and economic growth. *MZ Computing Journal, 5*(1) 1–5.

Stevens, E. (1995). *The grammar of the machine: Technical literacy and early industrial expansion in the United States.* Yale University Press.

Stoesz, B. M., & Eaton, S. E. (2022). Academic integrity policies of publicly funded universities in western Canada. *Educational Policy, 36*(6), 1529–1548.

Stokel-Walker, C. (2025, March 3). AI chatbots have telltale quirks: Researchers can spot them with 97% accuracy. *Fast Company*. https://www.fastcompany.com/91286162/ai-chatbots-have-telltale-quirks-researchers-can-spot-them-with-97-accuracy

Susilo, M. J., Sulisworo, D., & Beungacha, S. (2023). Technology and its impact on education. *Buletin Edukasi Indonesia*, *2*(02), 47–54.

Swanson, E. (2021). The case against the national breast implant registry. *Annals of Plastic Surgery*, *86*(3), 245–247.

Syed, F. M., & ES, F. K. (2023). Leveraging AI for HIPAA-compliant cloud security in healthcare. *Revista de Inteligencia Artificial en Medicina*, *14*(1), 461–484.

Taylor, J. S. (2023). Reassessing academic plagiarism. *Journal of Academic Ethics*, *22* (2), 211–230.

Teixeira da Silva, J. A., & Tsigaris, P. (2023). Human-and AI-based authorship: Principles and ethics. *Learned Publishing*, *36*(3), 453–462.

Vadakedath, S., Sudhakar, T., & Kandi, V. (2018). Assessment of conventional teaching technique in the era of medical education technology. *American Journal of Educational Research*, *6*(8), 1137–1140.

West, M. (2006, August). Cultural conceptions of intellectual property: The pirated disc market in Xi'An, China. *China*.

Winzenried, A., Dalgarno, B., & Tinkler, J. (2010). The interactive whiteboard: A transitional technology supporting diverse teaching practices. *Australasian Journal of Educational Technology*, *26*, 534–552.

Wisgalla, A., & Hasford, J. (2022). Four reasons why too many informed consents to clinical research are invalid: A critical analysis of current practices. *BMJ Open*, *12*(3), e050543.

CHAPTER 5

Creative Autonomy and Institutional Overreach

This chapter critically examines how institutional overreach, exemplified by stringent transparency demands from publishers and academic gatekeepers, poses significant threats to scholarly autonomy in the era of generative content creation. Despite calls for mandatory disclosure of artificial intelligence (AI)–assisted authorship — ostensibly to preserve integrity — this chapter reveals how transparency often becomes a tool for control, paradoxically diminishing trust and stifling innovation. Editors frequently reject disclosed AI-supported research despite recognizing its quality when authorship methods remain hidden, underscoring institutional resistance to technological change. Through considering historical parallels like the Scopes Trial and exploring disciplinary differences in citation practices, the chapter argues for pragmatic policies focused on evaluating scholarly outputs by their intellectual merit rather than the specifics of their creation. In proposing balanced guidelines, this chapter advocates safeguarding methodological privacy and creative autonomy, challenging the traditional publisher-driven paradigm to ensure academic freedom amid evolving technologies.

5.1 Navigating Power Dynamics in Publishing

Transparency, like honesty, is often touted as the best policy — but in academic publishing, it can be an oddly self-defeating virtue. Authors dutifully disclose their use of generative AI tools, expecting appreciation for their openness, only to discover they've handed editors a convenient reason for rejection. Indeed, experience reveals a curious paradox: Content praised for innovation when its AI origins remain undisclosed suddenly becomes problematic once transparency reveals its algorithmic roots. One editor candidly warned that openly admitting the use of generative tools would be tantamount to "selling one's career" — a sentiment oddly reminiscent of medieval guilds wary of those suspicious printing presses. Yet the solution to this dilemma isn't particularly revolutionary: simply evaluate scholarly outputs on their intellectual merit rather than the details of their creation. After all, traditional academia has long embraced a curiously similar method—careful cherry-picking from authoritative texts, direct quotation, and nuanced paraphrasing—to construct original arguments. Generative tools merely streamline this age-old practice, albeit without the hand cramps.

Research has demonstrated that human evaluations of content can be influenced by knowledge of its origin, particularly regarding AI-generated outputs. For instance, a study published in *ScienceDirect* examined how source disclosure affects the evaluation of AI-generated messages. The findings indicated that when participants were aware that content was AI-generated, their evaluations were less favorable compared to when they believed the content was human-authored, even if the AI-generated content was objectively superior (Lim & Schmälzle, 2024). Similarly, research highlighted in *MIT Sloan* explored perceptions of AI-created content. The study reported by Walsh (2023) found that participants exhibited a positive bias toward content they believed was human-created. Interestingly, when the source was undisclosed, AI-generated content was often preferred, suggesting that knowledge of AI involvement can negatively impact evaluations. In the realm of art, another study revealed that individuals tend to be negatively biased against AI-created artworks when aware of their origin. This bias diminishes when the source is unknown, indicating that human engagement in the creative process contributes positively to art appraisals (Bellaiche et al., 2023). Therefore, like the publishing world, these studies collectively suggest that human evaluations of content are significantly influenced by perceptions of its origin, with a discernible bias favoring human-generated work over AI-generated outputs, even when the latter may be of equal or superior quality.

At the same time, adding even more complexity, disciplinary cultures heavily shape attitudes toward authorship and citation, further coloring perceptions of AI assistance. The humanities, with notoriously lower citation frequencies, tend toward meticulous human-centered approaches, viewing labor-intensive authorship as indicative of scholarly rigor. Science and technology fields, conversely, lean into efficiency, readily embracing streamlined processes—including AI-generated summaries or analyses—that expedite research dissemination. Both positions hold merit, though neither should monopolize scholarly practice, especially given the rapid evolution of technology-enhanced research across all fields. The crux lies in recognizing that scholarly legitimacy stems from the quality of contributions rather than romanticizing the inefficiencies of traditional methods. Navigating these institutional power dynamics demands a candid reassessment: Transparency should serve clarity, not reinforce outdated biases or gatekeeping tactics cloaked in academic virtue.

5.1.1 Prompt and Circumstance: The Rise of Algorithmic Authorship

However, the integration of these new technologies into the academic publication lifecycle isn't exactly novel, but its burgeoning capabilities signal a radical departure from traditional methodologies, ushering in both remarkable efficiencies and notable complexities (Esplugas, 2023). Scholarly research, traditionally methodical and meticulous, has found an unexpected ally in AI, adept at handling segmented, repetitive tasks with mechanical precision (Zeichner, 1995). As researchers wade into the initial exploration of broad topics, these models shine brightly. Algorithms dedicated to topic modeling sift through expansive datasets, efficiently highlighting dominant themes and emergent trends across fields as disparate as population studies, microbiology, UI/UX design, and even the nuanced terrain of the humanities (Abdelrazek et al., 2023). These AI-driven insights enable scholars to pinpoint promising niches swiftly, bypassing weeks—or even months—of tedious manual literature scouring. Databases like PerplexityAI, Consensus, and Elicit further refine this process, offering personalized summaries and guiding researchers based on past interactions and contemporary academic trends (Heston & Khun, 2023).

Once a topic is narrowed down, researchers traditionally embark on the often-grueling quest for relevant monographs and scholarly articles—a task ripe for machine intervention. Leveraging natural language processing (NLP), literature discovery tools rapidly discern the relevance and context of vast bodies of text, expediting initial literature reviews (Wagner et al., 2022). The cumbersome chore of pinpointing specific chapters or sections within expansive works also succumbs gracefully to text mining and NLP, rapidly identifying precise references hidden within indexes, footnotes, and bibliographies (Robledo et al., 2023). Furthermore, AI automates the task of parsing bibliographies, swiftly assembling collections of focused studies and relevant research articles. Analytical platforms such as Scite.ai, capable of evaluating scholarly tone and context, distill diverse academic viewpoints, situating new research firmly within ongoing intellectual debates (Pinzolits, 2024). Equally invaluable are AI-enhanced citation management tools, which handle intricate formatting requirements with consistent precision, safeguarding scholarly integrity while significantly reducing clerical fatigue (Fui-Hoon Nah et al., 2023).

Yet even as AI demonstrates substantial potential in streamlining research preparation, the scholarly publication lifecycle continues to grapple with an enduring thorn: peer review. Despite its critical role in ensuring scholarly rigor, peer review remains dogged by substantial challenges, notably entrenched subjectivity (El-Guebaly et al., 2023). Editorial decisions are often far from impartial, shaped significantly by personal biases, institutional affiliations, and the notorious "Matthew effect"—the tendency to privilege already-established scholars, thereby reinforcing academic hierarchies and inequities (Berlin, 2023; Rigney, 2010). Further complicating matters, editorial authority frequently lacks transparency, overshadowed by opaque power dynamics and occasionally compromised by publishers' commercial interests (Teixeira da Silva et al., 2019).

Compounding these systemic issues is the chronic shortage of available reviewers. The weight of peer review often falls disproportionately upon a small cadre of experts, intensifying reviewer fatigue and potentially straining professional and personal relationships (Cheah & Piasecki, 2022). Such limitations risk reducing both the quality and diversity of reviews, with consequences compounded by intricate professional networks and personal biases among editors and reviewers (Huisman & Smits, 2017). Moreover, implicit racial biases present additional barriers, exemplified starkly in psychological sciences, where systemic inequalities persistently limit diverse perspectives in prestigious journals (Lee et al., 2013). Bias within peer review extends well beyond race, influenced significantly by authors' institutional prestige, nationality, gender, and discipline-specific preferences (Bancroft et al., 2022). These biases subtly yet substantially erode the fairness and reliability of peer review, emphasizing the critical need for increased impartiality in scholarly assessments.

Enter AI-assisted peer review, a promising disruptor of the entrenched scholarly status quo. AI-driven review processes introduce objectivity by adhering strictly to predefined evaluation criteria, effectively neutralizing human biases and prejudices (Checco et al., 2021; Javed et al., 2021). This impartiality ensures manuscripts are assessed purely on merit, enhancing fairness and consistency within the review process (Salah et al., 2023). AI's ability to swiftly process extensive textual data translates into rapid, consistent, and detailed feedback—qualities essential for dynamic, rapidly advancing research disciplines.

The latest generation of models substantially alleviates reviewer fatigue by automating preliminary manuscript evaluations. Submissions failing to meet baseline criteria are efficiently filtered, freeing human reviewers to engage deeply with intellectually challenging analyses (Cerdá-Alberich et al., 2023). These tools also further streamline reviewer selection by swiftly identifying experts whose interests align closely with submission topics, significantly enhancing process efficiency. A significant area of impact is increasingly becoming a hybrid of human and machine efforts—often referred to as a "cyborg" or "centaur" model. In this context, "cyborg" denotes a system where machine and human reviewers work in tandem, with generative tools handling preliminary assessments and humans providing nuanced evaluations. "Centaur" refers to a collaborative model where AI augments human capabilities, leading to more efficient and accurate reviews. AI can automate aspects such as plagiarism detection, consistency checks, and even suggest potential reviewers, thereby streamlining the process and reducing reviewer fatigue.

The future integration of this technology into academic publishing promises a radical and evolving transformation, accelerating the speed of research dissemination and enabling scholars to manage the exponentially increasing volume of publications across all disciplines. This can be contextualized within the exponential growth of data in recent years. It was estimated in 2024 that 90% of the world's data were created in just the past two years, reported by Bartley. Additionally, projections indicate that the total amount of data created, captured, copied, and consumed globally will reach over 180 zettabytes by 2025 (Michalowski, 2024). This rapid expansion underscores the accelerating pace of data generation in the digital age. Given the explosive growth in data is also scholarly output, meaning AI-powered tools will become indispensable, efficiently handling tasks from data synthesis to comprehensive literature surveys. Moreover, the peer-review process itself is inevitably shifting toward greater automation, where generative platforms rapidly handle initial screening, quality checks, and even complete evaluative tasks. In this evolving landscape, the human role in scholarship is progressively shifting toward that of director, editor, and curator, guiding and refining AI-generated content rather than producing every element firsthand. This shift is not speculative; it is already underway, marking a decisive move toward collaborative authorship where human scholars oversee and enhance computationally generated scholarship, emphasizing critical judgment, creativity, and intellectual stewardship.

As evidence of this shift, AI's role in content creation has reshaped the digital landscape dramatically. Since ChatGPT burst onto the scene in November 2022, AI-generated content online surged by an eye-watering 8,000% as of March 2024 (Mello, 2024). Clearly, digital creators have embraced their algorithmic muses enthusiastically—or perhaps desperately. Within the publishing sector, AI is rapidly becoming a standard tool rather than a curious novelty. A 2023 survey revealed that 23% of U.S. authors already rely on generative assistance, with 47% utilizing AI primarily for grammar corrections and another 29% employing it for generating fresh plot twists and character quirks (*ePublishing*, 2024). Human authors, it seems, are increasingly comfortable delegating creativity's grunt work to their digital counterparts.

The news media industry isn't exactly resisting this algorithmic charm offensive either. On a single day—July 1, 2024—AI-produced articles accounted for nearly 7% of total news output, equating to approximately 60,000 synthetic pieces hitting readers' feeds (Emi, 2024). The era of journalists hunched over keyboards might soon seem quaintly nostalgic. Liang et al. (2025) reinforce this narrative with a striking analysis spanning consumer complaints, corporate announcements, and even United Nations statements from January 2022 to September 2024. Their dataset is staggering: over 687,000 consumer complaints, 537,000 corporate press releases, nearly 16,000 UN statements, and—brace yourself—approximately 304 million job postings. Post-ChatGPT's meteoric launch, AI-generated content surged sharply. By late 2024, algorithms authored roughly 18% of consumer grievances, 24% of corporate communications, about 10% of small-business job advertisements (higher among younger companies, naturally), and an impressive 14% of UN statements. These aren't fringe experiments; automated text generation has firmly entrenched itself in our daily informational diet.

Looking forward, the acceleration of AI content production shows no sign of slowing. Some futurists estimate that by 2026, synthetic text could account for an astonishing 90% of all online material (Pereira, 2024). At this pace, human-crafted content might soon become a quaint artisanal specialty—enjoyed mainly by nostalgic bibliophiles and digital hipsters. Little wonder traditional institutions, long accustomed to controlling the narrative, are circling their wagons with wary eyes. Indeed, the nature of writing itself has quietly shifted over decades: Since Google's 1998 launch, we've been crafting content not primarily for human readers but for AI-driven algorithms parsing metadata and keywords. We've been writing for robots all along—whether we realized it or not.

5.1.2 Authority Anxiety: Institutions, Publishers, and the Battle over Creative Control

Debates over who holds ultimate authority in defining knowledge, creativity, and educational content are not new, echoing vividly in landmark moments such as the Scopes Trial of 1925. Known popularly as the "Monkey Trial," this infamous courtroom clash in Dayton, Tennessee, symbolized the profound struggle between institutional control and intellectual autonomy: traditional institutions and community beliefs pitted squarely against educators championing scientific innovation and academic freedom (Shapiro, 2016). Today, the rapid integration of generative AI tools into publishing and education has resurrected similar anxieties, reigniting the tussle between publishers, academic institutions, and authors over who precisely gets to determine legitimate methods of creation. On one side stand institutions and publishers, who assert their authority by imposing transparency mandates, method disclosure policies, and restrictive guidelines—often portraying these measures as necessary safeguards against ethical ambiguity and scholarly misconduct. On the other side are authors and creatives, increasingly viewing such mandates as instruments of institutional overreach, restricting intellectual exploration and curtailing innovation through bureaucratic micromanagement.

Much as the Scopes Trial hinged on who should determine educational curricula—the academy or the broader public—contemporary publishers and institutions grapple with the question of who sets the standards for creative methodologies in an AI-mediated world. Just as William Jennings Bryan argued passionately that educational content must align with prevailing societal and religious norms, many institutions today insist content creators must align their methods with established traditional academic processes. Conversely, modern-day authors and creators echo Clarence Darrow's appeals for intellectual openness, arguing that innovation arises from embracing new tools and methodologies—even (and especially) when they disrupt entrenched academic norms. As generative AI automates writing, analysis, and creative tasks at unprecedented scales, authors argue convincingly for the right to define how best to harness these emerging tools. This contention forces academia and publishing industries to revisit long-held assumptions about intellectual authority, requiring that educational and publishing policies strike a balance between necessary oversight and creative freedom.

At its core, the dispute reflects deeper anxieties about control, identity, and legitimacy: Publishers and institutions, fearing diminished authority, enforce increasingly rigid controls over disclosure and transparency; authors and creators, valuing intellectual independence and adaptability, resist what they perceive as an unwarranted intrusion into their creative process. Thus, the contemporary debate mirrors the historical tensions epitomized by the Scopes Trial—institutions assert their right to regulate content creation methods to preserve tradition and ensure conformity, while creators assert their autonomy to embrace disruptive innovation and intellectual evolution. Such tensions underscore the urgent need for dialogue and pragmatic policies that uphold standards of integrity and transparency without stifling the creative spirit crucial for advancing knowledge in an AI-driven future.

Educational authority has always been a battleground — today's tech-infused tug-of-war echoes vividly the intellectual melodrama of a century ago. The rapid-fire advancement of generative AI and other emerging technologies has reignited fierce debates over who should control educational content (Singha et al., 2024; Tuomi, 2024). Institutions, publishers, authors, and even the general public find themselves wrestling over the wheel, each asserting divergent visions of what education should achieve. Should curricula prioritize broad, interdisciplinary capabilities, fostering nimble thinkers for an uncertain world, or zoom in on precise, career-specific skills that promise immediate employability? This contest reflects broader societal struggles — politicians, educators, students, parents, and even the occasional meddling neighbor each stake their claims, shaping the educational agenda in their own ideological image. Today's clashes over the legitimacy of AI-generated content and methodologies echo this historic tension over authority, highlighting the stubborn resilience of questions about who decides what knowledge counts.

The Scopes Trial was a flashpoint precisely because it crystallized these enduring tensions into a gripping courtroom drama, broadcast across America. The case threw into stark relief the conflict between expert-driven curricula and populist educational control. John Scopes, a mild-mannered biology teacher, was prosecuted under the Butler Act for daring to teach human evolution in a publicly funded school, a subject outlawed in favor of creationist doctrine (Scopes & Bryan, 1925). His legal battle, championed by legendary lawyer Clarence Darrow against the fervent oratory of William Jennings Bryan, wasn't merely a spectacle; it symbolized a deeper ideological schism. Bryan, embodying traditional community norms and religious conservatism, championed populist authority over classroom content. Darrow defended the right — and indeed, the necessity — for educators to rely on expert scientific judgment rather than legislative mandate or public opinion. Scopes was convicted and fined $100, though his verdict was later overturned on a technicality — a narrative twist befitting such theatrical proceedings (Shapiro, 2016).

Yet the drama's conclusion settled nothing decisively, leaving an unresolved legacy of institutional anxiety about educational authority. Nearly a century later, the echoes of Scopes resonate through contemporary educational debates. The authority over curriculum today remains contested by powerful interest groups, each convinced of their moral and intellectual legitimacy. Controversies over teaching climate science, sex education, history, and, now, AI-generated content continue to mirror the dynamics of the Monkey Trial, pitting expert-driven scholarship against populist sentiment and institutional tradition (Žydžiūnaitė, 2016). Publishers and educational institutions, wary of losing control, often respond defensively, erecting rigorous transparency and disclosure mandates aimed at policing authorial methods. Such measures frequently reflect not genuine concerns for integrity but only the institutional anxiety about the democratization of knowledge creation—an anxiety vividly foreshadowed by the Scopes Trial. At its core, the conflict is about authority: Who decides how content is created, evaluated, and disseminated—the institutions and publishers traditionally gatekeeping knowledge, or the scholars, creatives, and educators producing it?

At the same time, the question is about control over knowledge creation and distribution between authors and publishers. This tension raises critical institutional questions: Should educators and publishers dictate the "acceptable" means of generating knowledge, narrowly prescribing methods that align with their historical norms, or should they empower authors, scholars, and creators to harness innovative AI-driven methods at their discretion? Emerging research underscores that as generative tools automate content production and analytical tasks, the premium placed on uniquely human capabilities—critical thought, ethical reasoning, and contextual adaptability—will only intensify (Noy & Zhang, 2023). Consequently, institutions face mounting pressure to either loosen their authoritative grip or risk obsolescence by enforcing restrictive guidelines. As publishers grapple with defining acceptable methodologies, the balance must shift toward supporting creative autonomy and intellectual adaptability, preparing students and authors alike to navigate ethical complexities and societal impacts in a future defined by generative AI (Preiksaitis & Rose, 2023). Thus, the question is less about technology itself and more about the distribution of power: Who gets to define legitimate scholarship and authentic creativity in the rapidly transforming landscape?

The traditional academic publishing system has long thrived on a paradox—scholars generously volunteer their intellectual labor, while publishers cheerfully rake in substantial profits. Academics don multiple hats as authors, reviewers, and editors, yet receive little beyond prestige or, occasionally, a complimentary subscription to a journal they themselves produce. Such benevolence fuels a lucrative industry, with leading publishers reporting annual revenues comfortably exceeding $10 billion, built almost entirely on unpaid scholarly efforts (Siler, 2017). Critics have aptly characterized this scenario as institutionalized exploitation, describing peer review as an "exploitation of the free labour of researchers" (Tennant, 2020). Ethical alarms ring loud here: Is it fair—or even rational—for scholars to subsidize a commercial enterprise from which they rarely reap any tangible rewards?

Unsurprisingly, tensions are reaching a boiling point. Legal battles have begun to chip away at the foundations of this peculiar arrangement, notably exemplified by recent class-action lawsuits accusing major publishers of antitrust violations directly tied to uncompensated reviewer labor (Uddin et al., 2024). Yet these courtroom dramas merely punctuate an ongoing, decades-long debate about fairness, equity, and the sustainability of the current academic publishing hegemony. This simmering unrest has sparked growing enthusiasm for alternative approaches, such as open-access publishing, which champions the radical idea that publicly funded research should actually be freely available to the public (Frank et al., 2023; Tennant et al., 2016; Willinsky, 2018).

As voices advocating for reform grow louder, the scholarly community increasingly recognizes that the current publishing landscape is not merely due for minor tweaks, but ripe for wholesale revolution. The ethos of transparency, democratization of knowledge, and fair compensation for intellectual contributions has moved from the periphery to mainstream discussion. It seems academia is finally ready to reconsider its relationship with commercial publishers, not as passive contributors to their profit margins but as active stakeholders demanding equitable recognition and fair treatment. After all, a model that endlessly profits from altruistic labor might be charmingly quaint—but sustainable and ethical, it certainly is not.

Long before generative AI became a contentious buzzword in artistic circles, independent artists and filmmakers raised their voices—often quite loudly—against the exclusivity and rigid gatekeeping perpetuated by art institutions and Hollywood's notorious closed ecosystem. These institutions, comfortably ensconced behind velvet ropes, decided who gained access to funding, galleries, distribution channels, and critical acclaim, leaving many independent creatives to peer in from the margins. Movements like Dada and Fluxus emerged precisely as rebellions against institutional elitism, poking fun at commercialism and snobbery with gleeful abandon. Fast-forwarding to the early 2000s, New York City's Antagonist Movement took up the mantle, staging provocative demonstrations, weekly art gatherings, and alternative exhibitions to lampoon the superficiality of the mainstream art world (Trumble & van Riemsdijk, 2016). Likewise, Fluxus artists of that era blurred boundaries between art and everyday life, advocating an inclusive, artist-centric creativity that thumbed its nose at the market-driven status quo (Higgins, 2002). Even more pointedly, Institutional Critique directly targeted the invisible power structures of museums, governments, and art markets, exposing the quiet machinations behind funding and curatorial decisions (Alberro &

Stimson, 2011; Sheikh, 2006). Similarly, Stuckism, peaking in the 2010s, openly protested the influence of heavyweight collectors and curators—most notably figures like Charles Saatchi—demanding a return to accessible, figurative art free from pretentious posturing (Harvey, 2012). These movements didn't just ask for change—they demanded artistic freedom from institutional meddling in matters of style, content, medium, and methodology.

Filmmakers outside Hollywood's gilded gates faced parallel challenges. For decades, the industry's centralized control over production, distribution, and financing limited the voices that could tell new stories, favoring well-connected insiders. Independent filmmakers like Latvian animator Gints Zilbalodis, creator of the distinctive film *Flow* (2024), persevered outside traditional studio pipelines, navigating financial struggles and logistical nightmares to deliver innovative storytelling (Brooks, 2025). Such creators consistently pushed back against studio-prescribed norms regarding narrative structure, visual style, and production workflow—often at great personal and financial risk.

Yet here's a curious twist of irony: Many artists who fervently rallied against institutional constraints on artistic freedom now find themselves among the staunchest critics of generative AI tools—tools offering precisely the sort of procedural liberation these same artists once championed. This pushback often stems less from ideological convictions than from anxieties about identity and professional relevance (as detailed in Chapter 1). Nevertheless, the logical conclusion of previous cries for artistic freedom and institutional reform aligns with the use of generative tools; these innovations directly challenge prescriptive, top-down mandates over artistic methods, mediums, and subjects. Indeed, the rapid evolution of these technologies may precipitate more than just a loosening of control; they may render these institutions irrelevant entirely. Thus, a provocative yet necessary question emerges: What can—or perhaps must—these entrenched institutions do to avert their own revolutionary dismantling? If history offers a lesson, it may well be that adaptation, genuine openness to methodological pluralism, and a willingness to embrace the rapidly shifting landscape of creativity could prove crucial to their survival.

5.2 Policies to Preserve Creative Responsibility and Autonomy

Given these mounting demands for institutional overhaul and the urgent calls for creative freedom, creative institutions now find themselves at a crossroads—adapt quickly, or risk obsolescence. With generative tools accelerating at dizzying speeds, it's clear that creativity is sprinting ahead, occasionally glancing back to wonder why institutions are still lumbering behind. Since the stable launch of ChatGPT 3.5 in February 2023, critical voices like Rosenberg (2023) have grown louder, highlighting academia's entrenched aversion to change — what he aptly terms "higher ed's ruinous resistance." Such resistance may feel comforting, like clinging to an outdated syllabus, but the reality outside the lecture hall demands more agility. Publishers and universities alike must reassess their policies, confronting head-on the tension between cherished traditions and the inevitable tides of technological innovation. It's time for an honest review: Rather than policing the methods behind creativity, institutions must learn to trust the final outcomes, much as educators judge a student's mastery by their work rather than their method of cramming. The ultimate test of creative output, after all, should remain the intrinsic quality, originality, and intellectual rigor of the work itself — not a bureaucratic scrutiny of how precisely that creativity was achieved.

5.2.1 Navigating the Maze: Current AI Policies in Publishing, News, and Academia

When AI first stepped onto the creative stage, reactions from publishers and academic institutions varied from wary curiosity to outright panic. Some saw an innovative partner; others saw the Frankenstein's monster — and swiftly grabbed their pitchforks. But as the initial hysteria subsided, clearer (and calmer) perspectives have begun to emerge. This section explores actual policies implemented by major academic publishers, popular media organizations, and universities, dissecting their differing stances on AI-generated research, journalism, creative writing, and classroom applications. From Elsevier's stringent academic declarations to Penguin Random House's cautious curiosity, we'll trace the complex boundaries these organizations have drawn. Universities, too, present a patchwork landscape, with disciplines such as scientific research, humanities, and creative writing each setting distinct rules of engagement. Finally, a comparative analysis will highlight key trends, contradictions, and ongoing flashpoints, illustrating why the path toward a coherent approach remains anything but straightforward.

When ChatGPT soared into popularity in early 2023, academic publishers were among the first to quickly scramble and set boundaries. Predictably, some responses bordered on hysteria: *Science*'s editor-in-chief Holden Thorp dramatically equated submitting AI-generated manuscripts to outright plagiarism, marking it as scientific misconduct unless explicitly approved by the journal (Park, 2023). *Nature* adopted a somewhat softer stance, permitting AI use if authors documented it clearly in methods or acknowledgments—though God forbid ChatGPT ever graces the byline (Nature, 2023). Elsevier, wielding significant clout with its 2,800 journals including heavyweight titles like *Cell* and *The Lancet*, similarly clarified that while AI can aid readability and language, human researchers must retain responsibility for data interpretation and scientific insights (Elsevier, 2023).

Major academic publishers have swiftly scrambled to set clear boundaries around AI's role in research and publishing. One consistent rule has emerged: AI tools simply cannot share credit as authors. Sorry, ChatGPT—no bylines for you, no matter how eloquent your prose. Accountability, after all, remains stubbornly human. Taylor & Francis formalized this stance in February 2023, notably the very month ChatGPT crossed the million-user threshold (Taylor & Francis, 2023). Publishers like Elsevier and Springer Nature have been equally explicit. Springer Nature's policy is straightforward: Large language models (LLMs), including Claude, Meta AI, Gemini, or ChatGPT, don't meet the criteria for authorship. Direct outputs from these models are thus off-limits for copyright under their name, reinforcing that authorship demands accountability — something an algorithm cannot yet shoulder. As the policy from Springer notes (https://www.springer.com/gp/editorial-policies/ artificial-intelligence--ai-/25428500?srsltid=AfmBOoo5ZyUjjviDUHYO1sc9X-UP0I73SEDRKyjAKpJiJmau_ZipaWssW#:~:text=Large%20Language%20 Models%20,and%20style%2C%20and%20to%20ensure):

> *Large Language Models (LLMs), such as ChatGPT, do not currently satisfy our authorship criteria (imprint editorial policy link). Notably an attribution of authorship carries with it accountability for the work, which cannot be effectively applied to LLMs. Use of an LLM should be properly documented in the Methods section (and if a Methods section is not available, in a suitable alternative part) of the manuscript. The use of an LLM (or other AI-tool) for "AI assisted copy editing" purposes does not need to be declared. In this context, we define the term "AI assisted copy editing" as AI-assisted improvements to human-generated texts for readability and style, and to ensure that the texts are free of errors in grammar, spelling, punctuation and tone. These AI-assisted improvements may include wording and formatting changes to the texts, but do not include generative editorial work and autonomous content creation. In all cases, there must be human accountability for the final version of the text and agreement from the authors that the edits reflect their original work.*

When it comes to generative AI images, publishers tread carefully, acknowledging the "fast-moving" landscape that changes almost daily. Rather than constantly rewriting their rules to keep pace with each new court decision or U.S. Copyright Office announcement, publishers prefer a cautious, yet practical, stance. Springer Nature's policy captures this sentiment clearly: It emphasizes adhering strictly to existing copyright laws and publication ethics, recognizing that generative AI imagery still navigates uncharted legal territory. In other words, they're staying firmly on the side of caution—keeping lawyers happy and avoiding courtroom drama. As the policy continues:

> The fast moving area of generative AI image creation has resulted in novel legal copyright and research integrity issues. As publishers, we strictly follow existing copyright law and best practices regarding publication ethics. While legal issues relating to AI-generated images and videos remain broadly unresolved, Springer Nature journals are unable to permit its use for publication.

Yet publishers recognize the need for a few exceptions, albeit under controlled conditions. For example, generative images sourced through trusted agencies with established contractual agreements—who presumably navigate legal complexities adeptly—are acceptable. Likewise, if the AI-generated visuals form part of the core subject matter in articles explicitly about AI, publishers will assess these submissions individually, and carefully, case-by-case. Furthermore, generative AI tools that rely on precise, verifiable scientific datasets—where accuracy, ethics, and copyright are demonstrably managed—may also earn approval. But there's a catch (isn't there always?): Publishers insist these exceptions be clearly labeled as "AI-generated" within the image itself, just in case readers mistake algorithmic creativity for human artistry.

In parallel, accountability isn't limited to authors alone; peer reviewers also shoulder significant responsibility. Publishers emphasize that, just as authors must stand behind the integrity of their manuscripts, reviewers must uphold rigorous standards when evaluating scholarly work. As Springer notes:

> Peer reviewers play a vital role in scientific publishing. Their expert evaluations and recommendations guide editors in their decisions and ensure that published research is valid, rigorous, and credible. Editors select peer reviewers primarily because of their in-depth knowledge of the subject matter or methods of the work they are asked to evaluate. This expertise is invaluable and irreplaceable.

> *Peer reviewers are accountable for the accuracy and views expressed in their reports, and the peer review process operates on a principle of mutual trust between authors, reviewers and editors. Despite rapid progress, generative AI tools have considerable limitations: they can lack up-to-date knowledge and may produce nonsensical, biased or false information. Manuscripts may also include sensitive or proprietary information that should not be shared outside the peer review process. For these reasons we ask that, while Springer Nature explores providing our peer reviewers with access to safe AI tools, peer reviewers do not upload manuscripts into generative AI tools.*

Ultimately, publishers are sending a clear, if slightly stern, message: Human accountability remains paramount. Authors and reviewers alike bear full responsibility for their manuscripts, critiques, and evaluations. Yes, AI tools can swoop in to polish grammar, finesse readability, and tidy up stylistic awkwardness—but they're strictly forbidden from usurping the core scholarly tasks. Drawing original conclusions, providing meaningful analysis, and exercising subject-matter expertise must firmly remain in human hands. Elsevier echoes this stance explicitly in their policy titled "The use of generative AI and AI-assisted technologies in writing for Elsevier," underscoring the boundaries placed on AI to preserve scholarly integrity (https://www.elsevier.com/about/policies-and-standards/the-use-of-generative-ai-and-ai-assisted-technologies-in-writing-for-elsevier#:~:text=Where%20authors%20use%20generative%20AI,the%20contents%20of%20the%20work) The policy states:

> *Where authors use generative AI and AI-assisted technologies in the writing process, these technologies should only be used to improve readability and language of the work and not to replace key authoring tasks such as producing scientific, pedagogic, or medical insights, drawing scientific conclusions, or providing clinical recommendations. Applying the technology should be done with human oversight and control and all work should be reviewed and edited carefully, because AI can generate authoritative-sounding output that can be incorrect, incomplete, or biased. The authors are ultimately responsible and accountable for the contents of the work.*

As such, transparency reigns supreme: "any use of generative AI must be disclosed," typically within the manuscript's Methods or Acknowledgments section. Publishers have made it clear—they're fine with a bit of algorithmic assistance, provided authors openly acknowledge the digital helping hand. As the policy continues:

> *Authors should disclose in their manuscript the use of generative AI and AI-assisted technologies and a statement will appear in the published work. Declaring the use of these technologies supports transparency and trust between authors, readers, reviewers, editors, and contributors and facilitates compliance with the terms of use of the relevant tool or technology. This policy is intended to cover new content creation only (i.e., new works or new content or chapters added to a revised work). Generative AI and AI-assisted technologies should not be used on previously published material.*

Again, ultimately responsibility rests squarely with the human author — not the AI. Elsevier explicitly emphasizes that LLMs or any other generative tools cannot be named as authors or coauthors, reinforcing that true accountability (and blame, for that matter) remains distinctly human. As they note:

> *Authors should not list generative AI and AI-assisted technologies as an author or co-author, nor cite AI as an author. Authorship implies responsibilities and tasks that can only be attributed to and performed by humans. Each (co-) author is accountable for ensuring that questions related to the accuracy or integrity of any part of the work are appropriately investigated and resolved and authorship requires the ability to approve the final version of the work and agree to its submission. Authors are also responsible for ensuring that the work is original, that the stated authors qualify for authorship, and the work does not infringe third party rights, and should familiarize themselves with Elsevier's Publishing Ethics policy before they submit.*

As with Springer, Elsevier enforces rigorous guidelines concerning AI-generated imagery, focusing heavily on preventing scientific obfuscation or misleading alterations. Their stance underscores a widespread concern within scholarly publishing about maintaining the integrity of visual data, given the potential for AI to introduce subtle, yet impactful distortions. Their stance underscores a widespread concern within scholarly publishing about maintaining the appearance of integrity of visual data, given the potential for AI to introduce subtle, yet impactful distortions. The explicit prohibition of modifications such as adding, removing, or significantly altering specific elements in scientific images ensures

Elsevier seems to preserve transparency, accuracy, and reproducibility—three cornerstones of credible scientific inquiry. As the policy states:

> Elsevier does not permit the use of generative AI or AI-assisted tools to create or alter images in submitted manuscripts. This may include enhancing, obscuring, moving, removing, or introducing a specific feature within an image or figure. Adjustments of brightness, contrast, or color balance are acceptable if they do not obscure or eliminate any information present in the original. Image forensics tools or specialized software might be applied to submitted manuscripts to identify suspected image irregularities.

However, Elsevier does allow AI tools when integrated directly into data acquisition or as part of the core research methodology. This allowance, though reasonable in theory, represents a rather precarious tightrope for researchers to walk. Scholars must carefully document their exact use of these AI-driven methods, clearly outlining software models, version numbers, and even manufacturers involved in data collection or analysis. While this seems straightforward enough on paper, the practical reality often feels like navigating a bureaucratic minefield one misstep, one vague explanation, and credibility could quickly unravel. This cautious flexibility indicates Elsevier's attempt to balance innovation with transparency, trusting researchers to maintain meticulous records of their AI-assisted processes. Still, for researchers juggling multiple demands, precisely delineating where human effort ends and algorithmic processing begins adds yet another layer of administrative complexity to their scholarly duties. For example:

> The only exception is if the use of generative AI or AI-assisted tools is part of the research design or research methods (such as in AI-assisted imaging approaches to generate or interpret the underlying research data, for example in the field of biomedical imaging). If this is done, such use must be described in a reproducible manner in the methods section. This should include an explanation of how the generative AI or AI-assisted tools were used in the image creation or alteration process, and the name of the model or tool, version and extension numbers, and manufacturer. Authors should adhere to the AI software's specific usage policies and ensure correct content attribution. Where applicable, authors could be asked to provide pre-AI-adjusted versions of images and/or the composite raw images used to create the final submitted versions, for editorial assessment.

Academic publishers have explicitly outlined boundaries for permissible AI assistance in scholarly writing. Elsevier, as previously noted, permits AI usage primarily for enhancing readability and grammatical precision under stringent human oversight. Yet, explicitly forbidden is the scenario in which an author merely approves original content or analysis that an AI independently generates. Exactly how journals expect authors to delineate the often subtle distinction between "AI-enhanced" and "AI-originated" content remains somewhat opaque—raising the question of how one might convincingly demonstrate their own intellectual fingerprints amid digital collaboration. Elsevier's policies emphasize the need to safeguard scholarly insights and the authors' critical thinking, though operationalizing these guidelines—especially given iterative interactions between researchers and generative AI tools—presents significant practical and epistemological challenges. What remains clear, though, is the publishers' uncompromising stance that the final responsibility rests firmly with human authors, who must diligently screen and correct any algorithmically introduced biases or inaccuracies. In other words, AI may write, but humans must vigilantly referee the final content.

Another contentious domain is the use of generative AI within peer-review and editorial processes, where restrictions are notably strict. Springer, for instance, expressly prohibits reviewers and editors from feeding manuscripts into generative AI systems, citing confidentiality and accuracy concerns. While publishers frame these prohibitions around data privacy and scholarly integrity, one must wonder whether such strictness also conveniently safeguards publishers' monetization strategies. Indeed, these restrictions reveal the industry's discomfort with the possibility of AI algorithms absorbing and internalizing proprietary knowledge—though realistically, the exact reproduction of specific manuscript content through an LLM is fundamentally impossible. Publishers thus seem driven less by genuine copyright concerns than by anxiety about losing control over their coveted content.

Despite these nuances, a general consensus is emerging across disciplinary boundaries: Human accountability remains paramount. Although initially spearheaded by science, technology, engineering, and mathematics (STEM)–oriented journals, humanities and social sciences have quickly aligned themselves with similar standards, reinforcing human oversight and explicit disclosure of AI involvement. Bioethics and humanities editors, for example, argue persuasively that AI lacks the capacity for moral responsibility and, thus, transparency in the form of disclosure remains nonnegotiable. Park (2023) highlights the Editors' Statement on Responsible Use of Generative AI Technologies in Scholarly

Publishing, emphasizing disclosure to editors, reviewers, and readers, along with clear explanations of AI's precise role in manuscript development. Yet beneath this explicit requirement lurks an implicit anxiety: Institutions and scholars alike worry that robots might trespass into intellectual terrain traditionally reserved for human experts. This fear, although often unstated, drives current policies and discourse—reflecting academia's underlying apprehension regarding the extent and implications of AI's encroachment into knowledge production.

By the same token, trade publishers, guardians of storytelling and human ingenuity, have approached the rise of generative AI with a wary caution—and perhaps just a touch of existential dread. Penguin Random House (PRH), for example, openly declares itself a staunch advocate of human creativity, insisting that no technology can replicate or replace genuine human imagination (https://www.penguin.co.uk/discover/articles/penguins-approach-to-generative-artificial-intelligence). This stance mirrors anxieties previously explored among creative writing students (as discussed in Chapter 1), revealing a common sentiment that creativity is strictly a human trait—not something to be algorithmically outsourced. PRH's commitment extends so far as to publicly promise that every book published under its banner will remain human-crafted from inception to print. In a further act of protective defiance, PRH now includes a stern notice on every copyright page explicitly prohibiting third-party AI training without consent—a move celebrated by groups like The Author's Guild, which advocate similar contract amendments to protect authors' intellectual property (*The Author's Guild*, 2024).

Yet not every publisher is entirely resistant to experimenting—cautiously—with AI. HarperCollins, ever the adventurous sibling, has taken measured steps into this controversial territory through a tightly controlled partnership with an AI firm. They struck a limited licensing agreement permitting AI training on select older nonfiction titles, but only with the explicit, opt-in consent of authors who are duly compensated (*Publishers Weekly*, 2024). The agreement carefully outlines how authors' original content can be used, placing strict limitations on AI outputs to ensure fairness and prevent wholesale appropriation. Such experiments indicate a nuanced approach emerging within the industry: Publishers are intrigued by AI's potential, yet deeply cautious about authorial rights, fair compensation, and preserving creative integrity.

Overall, the prevailing sentiment among trade book publishers remains firmly in defense of human creativity. AI may be permitted cautiously behind the scenes—enhancing market predictions, automating accessibility, or refining promotional strategies—but it is decisively excluded from authorship. Readers' trust that a book is born of authentic human experience and creativity remains sacred, an ideal publishers are determined to protect at nearly any cost. In short, trade publishers seem to view AI not as a potential new author but as a powerful assistant whose role must always be carefully managed — and whose creative ambitions must never overshadow those of the human writers whose stories they champion.

Moreover, news publishers, caught between embracing innovation and maintaining journalistic credibility, are rapidly crafting guidelines — often with a healthy dose of skepticism. Take *The New York Times* (*NYT*), for example: Ever cautious about diluting its esteemed journalistic standards, it permits AI tools primarily for mundane newsroom tasks like summarizing facts and assisting with copy editing (TipRanks, 2025). Humans, however, remain firmly in the editorial driver's seat. Journalists can delegate short summaries or promotional blurbs to their machine sidekicks, but the meat-and-potatoes reporting and writing require a distinctly human touch. Any snippet even slightly tainted by machine-generated prose must bear a clear label to alert readers, ensuring transparency remains paramount. Additionally, *NYT* explicitly bars staff from ethically murky uses of the tools — no deepfakes, no sneaky paywall bypasses, and absolutely no unmarked AI-generated images. This "trust but verify" approach keeps journalists fully accountable for every published word (sound familiar?).

Meanwhile, The *Washington Post* took things up a notch by appointing an actual "AI editor," tasked specifically with keeping their AI experiments ethical and error-free (Deck, 2024). The *Post*, keenly aware of embarrassing AI blunders elsewhere (such as CNET's notorious misfires with unchecked AI-generated articles), focuses heavily on accuracy and attribution. Their strategy? Pilot AI discreetly — assisting data journalism or spotting emerging trends — while ensuring all reader-facing AI content undergoes meticulous human scrutiny. It's a careful dance, balancing efficiency gains with the ever-present risk of misinformation.

Reuters, the global news powerhouse, has adopted a similarly cautious but clear-eyed policy, allowing journalists to integrate generative AI judiciously, provided transparency rules are followed religiously. *Reuters* explicitly mandates disclosure whenever content is predominantly AI-created (https://www.reuters.com/info-pages/reuters-and-ai/#:~:text=Reuters%20journalists%20at%20times%20use,AI%2C%20can%20be%20reported%20here). Practically, this mostly applies to the kind of automated financial or sports summaries readers might expect — but still, each piece must proudly declare its AI parentage. *Reuters* even extends this vigilance to its licensing partners, ensuring its stringent AI attribution standards are upheld downstream.

Across the news industry, recurring themes are clear: rigorous labeling of AI-generated material, human oversight at every juncture, and editorial accountability above all. The Associated Press (AP), among the first news organizations to adopt formal guidelines, bluntly advises journalists to verify everything and publish nothing AI-crafted without explicit review and disclosure — echoing *NYT*'s cautious stance. Audience trust remains fragile; readers show persistent discomfort with predominantly machine-written news (Dang, 2024). Hence, transparency isn't optional — it's survival. Interestingly, amid litigation involving the *NYT*, a judge highlighted a critical distinction: "facts" themselves cannot be copyrighted, only their interpretation. Meaning, ironically, LLMs trained on factual *NYT* data aren't infringing on intellectual property — a small consolation for anxious publishers.

As publishers and news organizations wrestle publicly with AI, behind the scenes, universities and colleges are quietly battling similar institutional anxieties. Most campuses have opted for a minimalist approach: Instead of crafting dedicated policies, they shoehorn generative AI use into preexisting academic integrity frameworks. Translation? Administrators neatly sidestep proactive decision-making, preferring to label unauthorized AI assistance as plain-old plagiarism or cheating. At the University of Texas, for example, the stance is crystal clear: No shiny new rules for ChatGPT — simply, "using AI-generated content without authorization or acknowledgment constitutes a violation of the university's honor code," just like copying from an uncredited source (University of Texas, 2023). Conveniently, this approach absolves administrators from explicitly defining acceptable uses of generative AI, placing the full responsibility squarely on students — and, by extension, on individual instructors crafting syllabus disclaimers. It's a classic bureaucratic dodge, maintaining the status quo under the thin veneer of upholding academic integrity. As their syllabus recommendation states (https://ctl.utexas.edu/chatgpt-and-generative-ai-tools-sample-syllabus-policy-statements#:~:text=Regarding%20the%20 potential%20use%20of,If%20adopted):

> *Regarding the potential use of generative AI tools, no changes in university policy are required. It is already a violation of policy for students to represent work they did not do as their own, and work generated by an AI system that is not credited to that system falls under that policy. While there are clear limitations on the use of these tools in certain contexts at UT, instructors have the discretion to explore them in the classroom. At present, the CTL recommends that UT faculty and instructors decide whether or not these tools fit within their pedagogical aims and clearly state their course policies in a designated section of their syllabi.*

Similarly, Columbia University positions unauthorized AI use squarely alongside traditional forms of cheating or plagiarism — same old wine, shiny new digital bottles. Their generative AI policy applies comprehensively to staff, faculty, students, and researchers, defining acceptable use clearly and sternly noting that it may evolve — so the community should regularly revisit it, ideally before stumbling into an academic minefield (https://provost.columbia.edu/content/ office-senior-vice-provost/ai-policy#:~:text=Generative%20AI%20Policy%20, Students%20are). In everyday terms, Columbia students must first obtain explicit instructor permission to deploy tools like ChatGPT, and, even then, they're expected to clearly disclose and cite any AI assistance within their submitted assignments. Likewise, Princeton University requires explicit confirmation of AI permissions from instructors and mandates clear disclosure of any generative AI assistance — an effort to prevent AI-generated content from slipping quietly into coursework, unnoticed and uncredited. As the policy states:

> As defined in section 2.4.7, generative artificial intelligence (AI) is not a source, since its output is not produced by a person. If generative AI is permitted by the instructor (for brainstorming, outlining, etc.), students must disclose its use rather than cite or acknowledge the use, since it is an algorithm rather than a source. Students are responsible for familiarizing themselves with and adhering to course and departmental policies regarding the use of generative AI. Inappropriate uses of the results of generative AI on any work submitted to fulfill an academic requirement, such as directly copying the output, representing output generated by or derived from generative AI as their own, exceeding the parameters specified by the instructor, or failing to disclose its use, would constitute violations of academic integrity.

These institutional guidelines reflect academia's enduring belief: Students' work should demonstrate their genuine understanding, effort, and creativity — with AI strictly a supporting player, not the lead actor. Yet this leads to an uncomfortable philosophical question: How exactly do institutions quantify "student understanding" in the age of AI? And where, precisely, do we draw the line between human cognition and algorithmic assistance? The clarity, or lack thereof, is ironically quite unclear.

Many universities encourage instructors to explicitly define AI-use policies within their syllabi. Harvard's undergraduate office, for example, "encourages all instructors to include a policy regarding the use and misuse of generative AI" in their course syllabi (https://oue.fas.harvard.edu/ai-guidance#:~:text=AI%20 Guidance%20%26%20FAQs%20,and%20misuse%20of%20generative%20

AI) (https://cte.ku.edu/ethical-use-ai-writing-assignments#:~:text=Ethical%20 use%20of%20AI%20in,view%20of%20AI%20becomes). This has resulted in varied responses: Some professors outright forbid AI involvement—AI being treated like an unwanted party guest—while others integrate it purposefully into the learning process. Predictably, disciplinary differences emerge. In coding or engineering courses, instructors might greenlight AI-assisted debugging or inspiration, provided students comprehend and acknowledge such aid. Conversely, in writing-intensive disciplines like philosophy or literature, instructors often prohibit AI-generated prose entirely, anxious to preserve students' authentic literary voices. Still, despite disciplinary discrepancies, universities universally agree on one principle: Failing to acknowledge AI-generated content is tantamount to academic dishonesty, as stated explicitly by institutions like the University of Kansas (https://cte.ku.edu/ethical-use-ai-writing-assignments#:~:text=Ethical%20use%20of%20AI%20in,view%20of%20AI%20becomes). Even where AI assistance is permitted, students typically must critically engage with the outputs, shaping raw AI suggestions into meaningful academic work.

Interestingly, these policies can generally be grouped into three intuitive categories—think of it as academia's traffic-light system: Red for "no AI allowed"; yellow for "limited, supervised use"; and green for "open season." However, research indicates faculty often neglect including these policies explicitly in syllabi, erroneously assuming their stance on AI is universally understood. The predictable result? Confusion during disciplinary proceedings when academic affairs departments find themselves unable to enforce implied prohibitions lacking written confirmation. Compounding this issue, instructors frequently fail to specify exactly when, how, or for which assignments AI assistance is permissible. Vague statements like "AI usage permitted only as explicitly stated by the instructor" leave students adrift in ambiguity. Perhaps most glaringly, these policies often fail to acknowledge the reality of contemporary writing practices discussed extensively in Chapter 3. In the world of generative text, there are no neat boundaries—no simple point where human contribution clearly stops and machine assistance begins. Writing today is fundamentally cyborg-like, seamlessly blending human insight with algorithmic output in a way traditional academic policies rarely capture.

Beyond the classroom, universities are drafting guidelines governing faculty and researchers' use of generative AI, echoing publishers' cautious approach. Researchers are advised to explicitly document AI usage when preparing manuscripts or analyzing data, and certainly not to credit AI as an author—responsibility, after all, remains decidedly human, as noted clearly by institutions like the University of Texas at Arlington (https://resources.uta.edu/research/policies-and-procedures/ generative-artificial-intelligence.php#:~:text=Use%20of%20Generative%20Ar- tificial%20Intelligence,places%20and%20manners%20of%20use). Additionally, institutions emphasize strict handling of confidential and sensitive data. Faculty members are discouraged from casually inputting proprietary research findings or student information into third-party AI systems, fearing privacy breaches and data security risks. To navigate these evolving complexities, some universities, such as the University of Minnesota, have established specialized task forces examining AI's impact on pedagogy and scholarship.

Moreover, some institutions have issued comprehensive best-practice guide- lines to promote ethical AI use across campus. The University of Maryland's guidelines, for example, underscore human oversight, transparency, and equity, urging faculty and students to maintain ownership and accountability for any AI-generated output (https://ai.umd.edu/resources/guidelines#:~:text=The%20 University%20of%20Maryland%20,their%20teaching%2C%20learning%2C%20 and%20work). They emphasize compliance with privacy laws, particularly FERPA (Family Educational Rights and Privacy Act), and advocate equitable access to AI technologies—ensuring no student is unfairly advantaged or disadvantaged by AI integration. In essence, universities seek equilibrium: harnessing AI's educational potential without sacrificing essential academic values. While en- couraging students to learn how AI can support creativity, ideation, and analysis, institutions simultaneously reinforce the primacy of critical thinking, originality, and ethical rigor—qualities that no amount of AI automation should supplant.

Across academic publishing, commercial media, and educational institutions, distinct yet overlapping patterns have emerged in AI usage policies—patterns that reveal deeper anxieties and cautious optimism regarding AI's expanding role. One universal agreement across all sectors is human accountability. AI systems, despite their impressive capabilities, cannot bear moral or legal responsibility (no courtroom is prepared to prosecute ChatGPT). Academic journals, from Springer to Elsevier, firmly declare AI-generated text cannot earn authorship status; only humans hold responsibility for published content's integrity. Penguin Random House takes a similar stance in commercial publishing, vigorously defending human creativity—lest readers suspect their favorite novel was conjured entirely by algorithms. News outlets, ever mindful of their credibility, echo these sen- timents. Even when AI assists reporters, human oversight remains sacrosanct. After all, no journalist wants their Pulitzer awarded to GPT-5.

Transparency and disclosure also consistently appear in policies as nonnegotiable principles. Whether it's an academic manuscript, a news report, or a student essay, disclosure of assistance is uniformly mandated. Springer and Elsevier journals require authors to clearly document use, typically in methods or acknowledgments sections — omission risks outright rejection. Similarly, universities mandate students cite AI as they would any other source, ensuring academic honesty. Meanwhile, media outlets like *The NYT* and *Reuters* carefully label generated segments of content. This transparent stance serves dual purposes: Ethically, it prevents deception; practically, it warns audiences — and wary editors — to apply additional scrutiny to content potentially riddled with subtle hallucinations.

Permitted and prohibited AI uses vary by sector, reflecting differing institutional objectives. Academic publishers typically allow for linguistic polishing or graphics but prohibit generative tools from independently producing research conclusions or engaging in peer review. Newsrooms cautiously permit assistance for routine summaries and data processing, but investigative journalism and sensitive stories remain steadfastly human territory. Trade publishers tread lightly, exploring use primarily behind the scenes — for market analytics or experimental translations — but generally draw the line at publishing AI-authored manuscripts. Educational institutions present the most diverse approaches: Some ban the technology entirely in certain courses, prioritizing authentic student creation, while others incorporate it openly into curricula to cultivate digital literacy. Each sector, unsurprisingly, calibrates integration to preserve its core values: scholarly originality, journalistic truth, literary creativity, and educational integrity.

Ethical concerns — accuracy, bias, and fairness — permeate these policy frameworks, underscoring widespread caution. Publishers frequently highlight the uncanny ability to authoritatively produce plausible yet erroneous or biased statements, necessitating rigorous human vetting. News media, having witnessed embarrassing AI-generated missteps (e.g., CNET's high-profile errors), established stringent oversight measures to safeguard journalistic integrity. Academic institutions emphasize fairness in student evaluation: if assistance is allowed, instructors must ensure equitable grading between machine users and traditional learners, thus preventing inadvertent advantage or disadvantage. Confidentiality, too, remains critical; uploading sensitive documents to public smart platforms is explicitly prohibited, lest proprietary information becomes an accidental viral sensation.

Amid these ongoing negotiations, policies are evolving documents — always in beta, constantly adjusted in response to technological advances. Cross-sector collaborations have surfaced, notably exemplified by publishers and universities sharing guidelines and best practices. The International Committee of Medical Journal Editors, for instance, released recommendations widely adopted across scientific publishing, while universities like Maryland have proactively developed comprehensive frameworks promoting "ethical" integration. Institutions also recognize positive potentials: Wiley's guidelines encourage authors to enhance linguistic clarity through AI without sacrificing their unique scholarly voice. Likewise, educational institutions acknowledge potential to democratize learning—provided it supplements rather than replaces critical thinking.

Comparatively, academia tends to frame policies around integrity and ethics, stressing originality and scholarly rigor. Conversely, commercial publishers and news media prioritize consumer trust, copyright protection, and legal compliance, keenly aware that their audiences demand human authenticity. Yet, despite sector-specific nuances, consensus crystallizes around one central tenet: AI is a supportive tool—not an autonomous creator. Humans remain firmly in the driver's seat, responsible for steering intellectual and creative outputs through labyrinthine possibilities. In the coming years, these cautious guidelines will undoubtedly evolve. Stakeholders across all sectors must continuously recalibrate their standards of responsible use, balancing innovation with accountability, efficiency with authenticity, and technological novelty with timeless human values.

5.2.2 From Restriction to Responsibility: Rethinking Legal Boundaries

Creative industries are caught in a curious dance with AI—unsure whether to embrace their partner or keep a wary distance to avoid stepping on ethical or professional toes. Initially, institutions responded with protective instincts, banning or sharply restricting usage amid fears of plagiarism, misinformation, or even job displacement. Yet, outright prohibitions quickly proved not just impractical but also creatively stifling. A new consensus has emerged, advocating clear yet flexible frameworks that permit AI to flourish as a collaborator rather than casting it as a villain. This approach frees creators to leverage tools, enriching human imagination rather than supplanting it. Copyright laws have slowly adapted to this reality. In Europe, explicit exceptions for text and data mining (TDM) have opened the gates for training, recognizing that without extensive datasets, generative tools become little more than a toy car without batteries (Varese & Battistella, 2023). The U.S. fair-use doctrine, validated notably by the *Authors Guild v. Google* case, similarly supports transformative uses, allowing models to legally ingest massive amounts of data from existing works (Dilworth, 2024).

This ensures that creators can harness it without constantly fearing litigation, framing it as just another powerful analytical instrument rather than some intellectual pirate.

Meanwhile, the realm of intellectual property is undergoing a quiet revolution. The U.S. Copyright Office recently clarified that works produced with significant human oversight and creative direction—even when AI-driven—can still enjoy copyright protection (U.S. Copyright Office, 2025). In other words, the human guiding hand remains pivotal; pure machine-generated content alone remains orphaned, unprotected by copyright law. The United Kingdom has gone further, generously assigning authorship rights of purely AI-generated works to the person who initiated and guided their creation, even if no traditional human "author" exists (Blaszcyk, 2023). Such measures reflect a pragmatic acknowledgment: If human creativity guides the process, legal protections should naturally follow. Patents, too, are being reexamined. Traditional laws stubbornly insist on human inventors, but industry experts now advocate reforms to accommodate inventions where AI significantly contributes (Dilworth, 2024). Proposals range from creating special categories for "AI-assisted inventions" to redefining inventorship itself. By recognizing the extent of the role played by AI in research and innovation that led to a usable, verifiable, or authentic outcome, legal systems can preserve innovation instead of invalidating valuable creations merely because algorithms played a substantial role.

Underlying all these changes is a recognition that freedom of creative expression inherently includes the right to explore and experiment with emerging tools like AI. Just as no one outlawed digital photography when it replaced film, courts and policymakers are increasingly reluctant to curb creativity through restrictive rules. Instead, they prefer viewing it as another tool in the creative toolbox, updating interpretations and guidelines to accommodate technological innovation. Thus, creators using smart platforms can confidently experiment without the threat of losing legal protections or having their work unfairly categorized as plagiarism or infringement. These evolving legal frameworks underscore a critical shift in thinking: Instead of instinctively erecting fences against AI-driven creativity, society is gradually dismantling barriers. Through clarifying authorship, affirming fair use, and redefining invention rights, these policies liberate human creators to harness such potential. Rather than stifling creativity with fears of machine capabilities, they empower artists and innovators to safely explore new frontiers, ensuring the creative process remains dynamic, expansive, and distinctly human-led.

Rather than enforcing draconian bans, institutions and publishers can foster creative growth by crafting balanced policies that guide, rather than restrict, usage.

A crucial starting point is clarifying authorship criteria. Academic publishers like Elsevier, Science, and IEEE have sensibly determined that smart systems, lacking conscience and coffee breaks, cannot assume authorship or accountability. Humans alone shoulder the creative and ethical responsibility. By making this policy explicit, publishers and institutions erase ambiguity, ensuring that it serves as collaborator, not surrogate creator. Thus, institutions must go beyond mere policy declarations by actively providing guidance and training. Simply prohibiting or allowing AI without instruction is akin to handing someone a violin and expecting Beethoven without lessons—unlikely, messy, and potentially painful. Instead, organizations should emulate publisher Wiley, which has issued practical guidelines helping authors enhance their writing with AI without losing their authentic voice. Universities could similarly offer courses or workshops that teach students how to responsibly use AI tools, verify generated content, mitigate bias, and properly attribute their work to their digital co-creators. Education empowers users, transforming potential misuse into productive mastery.

Moreover, institutions should foster a culture of ethical experimentation rather than cautious avoidance. Leaders must communicate clearly that AI-assisted creativity is welcomed and encouraged. For example, *Cosmopolitan* magazine bravely piloted a generated cover, explicitly stating the technology was "just another graphic software tool among many"—not a robot usurper of creative control (Liu, 2022). Such pilot projects demonstrate institutional commitment to innovation and normalizes AI's role as just another paintbrush in the creative toolkit. Simultaneously, thoughtful institutions and publishers must establish robust policies safeguarding intellectual property and privacy. Clear contractual terms should specify that human creators (or their organizations) maintain copyright over AI-assisted works, avoiding ambiguity over who actually "owns" the output. Publishers should require explicit permission before using proprietary material—manuscripts, datasets, or content—for AI training, protecting both privacy and intellectual investments. In the entertainment sector, the Writers Guild's recent agreements explicitly prevent AI from diluting writers' credits or financial rewards, illustrating how to balance technological innovation with respect for creative rights (Scherer, 2023).

However, even the most supportive policies and progressive laws can't guarantee widespread acceptance if creative communities remain skeptical about AI. Bridging that cultural gap requires strategic moves to normalize use—not just tolerate it grudgingly. A crucial step involves embedding literacy into educational frameworks across schools, universities, and creative training programs. Rather than simply cautioning students against potential misuse, educators should integrate the tools actively into the curriculum. But education alone isn't enough. Public perceptions shift significantly when people witness respected creators using AI successfully. Highlighting such success stories provides tangible proof of AI's potential to amplify creativity. Visual artists employing generative models to craft entirely new aesthetics or filmmakers leveraging AI for groundbreaking special effects exemplify this potential (Mazzone & Elgammal, 2019). Promoting these narratives through conferences, workshops, and online communities can debunk misconceptions of AI-assisted work as somehow lesser or "cheating." Positive examples normalize the technology, framing AI not as a shortcut, but as a legitimate creative choice expanding artistic horizons.

Beyond visibility, fostering collaborative communities around AI further dismantles resistance. Hackathons, workshops, or shared experimental projects—like musicians collectively developing AI-generated compositions—establish inclusive spaces for creators to explore AI together (Sturm & Ben-Tal, 2021). These engagements create a culture of cooperative creativity, transforming AI from an existential threat into a collaborative partner. Moreover, involving diverse creators in AI's development ensures that evolving norms and practices reflect broad perspectives, making the technology accessible and appealing to a wider creative community.

Yet acceptance demands addressing legitimate concerns openly—fears of plagiarism, loss of originality, or job displacement aren't merely irrational anxieties. Rather than dismissing these worries, clear guidelines and assurances must be communicated. Platforms now routinely prohibit generating art in the style of living artists without consent, addressing concerns about artistic autonomy and appropriation head-on. Transparency measures like provenance tracking for AI-generated works or clear honor codes for students reduce fears, building trust through explicit ethical standards (Floridi & Chiriatti, 2020). When creators understand that guardrails exist to protect their originality and employment, acceptance becomes far more likely. Gradual, iterative implementation also smooths acceptance. Institutions introducing AI incrementally—for instance, first permitting AI in preliminary brainstorming, then expanding its use as users gain comfort—can significantly ease resistance. Publishers might pilot limited "AI-assisted editing" projects before broader adoption, reflecting creators' comfort levels and feedback. Over time, small successes from pilot programs validate AI's usefulness, transforming initial skeptics into enthusiastic adopters.

At the core, effective communication remains paramount. Creative communities should be actively involved in ongoing conversations about AI's role—not as gatekeepers to prevent innovation but as stakeholders shaping responsible adoption. A shared understanding that AI, much like Photoshop or computer-generated imagery (CGI) before it, is merely another tool to realize artistic visions helps normalize its usage, making it an integral rather than controversial part of creative workflows. Analyzing acceptance across industries reveals distinct patterns reflective of each field's unique priorities. Academic publishing initially reacted strongly to its emergence, with high-impact journals like *Science* labeling uncredited generated text as misconduct. Conversely, *Nature* and JAMA Network opted for conditional acceptance, permitting clearly disclosed assistance while reinforcing human accountability. Similarly, universities transitioned from viewing AI-generated content as outright plagiarism to developing detailed policies requiring transparency and proper citation. Although the pace varies, academia has collectively moved toward careful integration rather than blanket prohibition.

The visual arts faced more pronounced controversies—like the infamous 2022 incident where an AI-generated piece secured a blue-ribbon prize, igniting heated debates over artistic authenticity (Brittain, 2024). These tensions led some galleries and competitions to initially ban generated works. Yet, increasingly, art communities have created separate categories or guidelines, allowing assisted creations while preserving traditional values of originality. This measured acceptance acknowledges the creative potential without entirely discarding conventional artistic standards. Likewise, music presents an even more complex relationship. While established music labels aggressively combat unauthorized AI-generated tracks mimicking famous artists, independent musicians embrace the future more willingly. Notably, artist Grimes publicly invited creators to use an AI version of her voice, splitting royalties evenly—a radical yet pragmatic approach reframing AI as collaboration and not as competition (Gendron, 2023). Thus, music industry acceptance is growing cautiously, driven by ethical frameworks protecting original creators and ensuring fair use, yet increasingly comfortable with it as a creative partner. Film and media industries have perhaps integrated the tech most extensively, thanks to Hollywood's historical affinity for technological innovation. Yet this enthusiastic adoption sparked critical labor debates, exemplified by the 2023 Writers Guild negotiations. The resulting groundbreaking agreement explicitly permitted writers to utilize AI tools voluntarily, simultaneously prohibiting studios from using AI to undermine writers' credits or compensation (Scherer, 2023). Such negotiated policies reveal the industry's balanced stance—embracing the efficiency in technical processes while firmly safeguarding human creative roles and contributions.

Comparing these industries highlights both divergent approaches and commonalities. Academia emphasizes intellectual integrity, transparency, and proper attribution, while the arts grapple with authenticity, originality, and fair compensation. Music and film industries blend creativity with practical protections for creators, reflecting industry-specific tensions. Yet, across all fields, an overarching shift is evident: Initial resistance gives way to acceptance once ethical policies, transparency, and clear guidelines emerge. In the end, the goal of these legal frameworks, institutional policies, educational strategies, and industry-specific examples converges around a singular vision: enabling creators to freely utilize AI without undue restrictions. Comprehensive best practices underscore that outright bans stifle innovation; embracing AI responsibly through acceptable standards promotes both creative freedom and accountability. Legal systems increasingly provide flexibility through copyright exceptions and clarified authorship definitions. Educational efforts cultivate literacy, while industry policies ensure fairness, quality control, and intellectual property protection. Practically, institutions and creators alike should update guidelines explicitly permitting AI-assisted work and safeguarding human ownership of the creative process. Pilot projects and educational initiatives can reshape cultural attitudes, moving communities beyond fear toward responsible acceptance. This balanced integration benefits everyone—empowering creators with innovative tools, enriching audiences with novel experiences, and maintaining the ethical standards critical to creative integrity.

Throughout this chapter, one essential truth has become clear: Generative integration into creative, academic, and publishing spheres is both inevitable and transformative. Across these varied landscapes, from university classrooms and scholarly journals to film studios and media houses, policies have rapidly evolved—sometimes thoughtfully, sometimes impulsively—to regulate this new creative frontier. Publishers like Elsevier and Springer Nature have drawn bright lines, clearly stating that these tools may assist but must never replace human accountability. Trade publishers like Penguin Random House underscore the centrality of human creativity, vowing to protect authors' rights against algorithmic encroachment—ironically, by employing AI behind the scenes for marketing and forecasting. News media such as *The NYT* have adopted cautiously optimistic approaches, harnessing efficiencies for routine tasks but reinforcing human editorial oversight for content integrity. Academia, meanwhile, remains torn: Some institutions enthusiastically incorporate AI literacy into curricula, while others clutch honor codes tightly, wary of "unauthorized assistance."

Yet the landscape is uneven, as each sector navigates its distinct priorities and anxieties. Academia's chief concern remains intellectual integrity and authenticity of thought; creative industries fret over originality, authorship, and job security; news organizations vigilantly safeguard public trust and factual accuracy. All these domains have converged on the most important common denominator: explicit human accountability for content that is copyrightable. The emphasis across industries is currently (and unfortunately) on transparency, integrity, and maintaining audience or consumer trust, with humans firmly at the wheel. Only one of these is realistic. However, formal policies alone do not dictate professional acceptance. While a journal or publisher might explicitly permit the use of generative tools, editorial boards, reviewers, or academics within specific disciplines often quietly establish contrary professional norms—informal, unspoken rules shaping what is genuinely acceptable in practice. In other words, despite official guidelines allowing assistance, peer reviewers or editors may tacitly reject generated contributions, driven by lingering skepticism or philosophical discomfort. This discrepancy between policy and professional culture can become a subtle form of gatekeeping: officially sanctioned yet practically forbidden.

Such tensions reflect broader, historic struggles for creative autonomy. At heart, this book argues not that all creators should embrace generative AI—but rather that each creator must retain the freedom to make that choice independently. The existence of differing professional norms is acceptable—indeed, even healthy—as long as these remain voluntary conventions, reflective of collective choice rather than imposed doctrine. Just as religious converts may rightfully choose their path but falter ethically when coercing others, creative professionals cross an ethical line when they attempt to impose their norms on colleagues who do not share their convictions. Freewill in creativity must remain sacrosanct, defended against ideological conformity, even in an era where generative tech disrupts deeply ingrained beliefs about authorship, originality, and intellectual labor.

The goal then is balance: Institutions and industries should continue refining transparent, responsible policies, simultaneously respecting the diverse, self-determined norms within creative communities. Only by preserving this delicate equilibrium between organizational regulation and individual autonomy can creators truly flourish—free to adopt or reject AI as suits their personal and professional convictions. Creativity thrives best not through rigid conformity, but in a vibrant ecosystem where varied methods coexist, challenging and enriching one another in perpetual, dynamic dialogue.

References

Abdelrazek, A., Eid, Y., Gawish, E., Medhat, W., & Hassan, A. (2023). Topic modeling algorithms and applications: A survey. *Information Systems*, *112*, 102131.

Alberro, A., & Stimson, B. (Eds.). (2011). *Institutional critique: An anthology of artists' writings*. MIT Press.

Bancroft, S. F., Ryoo, K., & Miles, M. (2022). Promoting equity in the peer review process of journal publication. *Science Education*, *106*(5), 1232–1248.

Bartley, K. (2024, December 11). Big data statistics: How much data is there in the world? *Rivery*. https://rivery.io/blog/big-data-statistics-how-much-data-is-there-in-the-world/

Bellaiche, L., Shahi, R., Turpin, M. H., Ragnhildstveit, A., Sprockett, S., Barr, N., & Seli, P. (2023). Humans versus AI: Whether and why we prefer human-created compared to AI-created artwork. *Cognitive Research: Principles and Implications*, *8*(1), 42.

Berlin, S. (2023). Reconsidering editorial consideration: Changing editorial assessment could reduce subjectivity in the publication process. *EMBO Reports*, *24*(11), e58127.

Blaszcyk, M. (2023, November 6). Contradictions of computer-generated works' protection. Wolters Kluwer. https://copyrightblog.kluweriplaw.com/2023/11/06/contradictions-of-computer-generated-works-protection/#:~:text=,%E2%80%9D

Brittain, B. (2024, September 26). Artist sues after US rejects copyright for AI-generated image. *Reuters*. https://www.reuters.com/legal/litigation/artist-sues-after-us-rejects-copyright-ai-generated-image-2024-09-26/

Brooks, X. (2025, March 9) Halt Disney! Flow's director, and fellow upstart animators, on a new era for the artform. *Guardian*. https://www.theguardian.com/film/2025/mar/09/animation-oscars-flow-memoir-of-a-snail-wander-to-wonder

Cerdá-Alberich, L., Solana, J., Mallol, P., Ribas, G., García-Junco, M., Alberich-Bayarri, A., & Marti-Bonmati, L. (2023). MAIC–10 brief quality checklist for publications using artificial intelligence and medical images. *Insights into Imaging*, *14*(1), 11.

Cheah, P. Y., & Piasecki, J. (2022). Should peer reviewers be paid to review academic papers? *The Lancet*, *399*(10335), 1601.

Checco, A., Bracciale, L., Loreti, P., Pinfield, S., & Bianchi, G. (2021). AI-assisted peer review. *Humanities and Social Sciences Communications*, *8*(1), 1–11.

Dang, S. (2024, June 17). Global audiences suspicious of AI-powered newsrooms, report finds. *Reuters*. https://www.reuters.com/technology/artificial-intelligence/global-audiences-suspicious-ai-powered-newsrooms-report-finds-2024-06-16/#:~:text=Global%20audiences%20suspicious%20of%20AI,news%20produced%20mostly%20with%20AI

Deck, A. (2024, March 28). The *Washington Post*'s first AI strategy editor talks LLMs in the newsroom. *NiemanLab*. https://www.niemanlab.org/2024/03/the-washington-posts-first-ai-strategy-editor-talks-llms-in-the-newsroom/#:~:text=At%20the%20Post%2C%20Connelly%E2%80%99s%20job%2C,levels%2C%20while%20maintaining%20editorial%20standards

Dilworth, M. (2024, March 4). AI & intellectual property: Artificial intelligence legal implications. *Dilworth IP*. https://www.dilworthip.com/resources/news/artificial-intelligence-and-intellectual-property-legal-issues/#:~:text=notable%20example%20is%20Authors%20Guild,precedent%20in%20the%20legal%20landscape

El-Guebaly, N., Foster, J., Bahji, A., & Hellman, M. (2023). The critical role of peer reviewers: Challenges and future steps. *Nordic Studies on Alcohol and Drugs*, *40*(1), 14–21.

Emi, B. (2024, August 8). 60,000 AI-generated news articles are published every day. *Newscatcher*. https://www.newscatcherapi.com/blog/60-000-ai-generated-news-articles-are-published-every-day

ePublishing. (2024, December 5). AI in publishing [revolutionizing the future of content creation and compliance]. https://www.epublishing.com/news/2024/dec/05/ai-publishing/

Esplugas, M. (2023). The use of artificial intelligence (AI) to enhance academic communication, education and research: A balanced approach. *Journal of Hand Surgery* (European Volume), 48(8), 819–822.

Floridi, L., & Chiriatti, M. (2020). GPT-3: Its nature, scope, limits, and consequences. *Minds and Machines*, *30*, 681–694.

Frank, J., Foster, R., & Pagliari, C. (2023). Open access publishing–noble intention, flawed reality. *Social Science & Medicine*, *317*, 115592.

Fui-Hoon Nah, F., Zheng, R., Cai, J., Siau, K., & Chen, L. (2023). Generative AI and ChatGPT: Applications, challenges, and AI-human collaboration. *Journal of Information Technology Case and Application Research*, *25*(3), 277–304.

Gendron, W. (2023, April 24). Grimes says she'll split the royalties on any "successful" AI-generated song that uses her voice. *Business Insider*. https://www.businessinsider.com/grimes-split-royalties-with-ai-generated-music-using-her-voice-2023-4#:~:text=While%20the%20music%20industry%20reckons,technology%20might%20have%20to%20offer

Harvey, P. A. (2012). *Stuckism, punk attitude and fine art practice: Parallels and similarities*. University of Northumbria at Newcastle.

Heston, T. F., & Khun, C. (2023). Prompt engineering in medical education. *International Medical Education, 2*(3), 198–205.

Higgins, H. (2002). *Fluxus experience*. University of California Press.

Huisman, J., & Smits, J. (2017). Duration and quality of the peer review process: The author's perspective. *Scientometrics, 113*(1), 633–650.

Javed, S., Adewumi, T. P., Liwicki, F. S., & Liwicki, M. (2021). Understanding the role of objectivity in machine learning and research evaluation. *Philosophies, 6*(1), 22.

Lee, C. J., Sugimoto, C. R., Zhang, G., & Cronin, B. (2013). Bias in peer review. *Journal of the American Society for Information Science and Technology, 64*(1), 2–17.

Liang, W., Zhang, Y., Codreanu, M., Wang, J., Cao, H., & Zou, J. (2025). The widespread adoption of large language model-assisted writing across society. *arXiv preprint arXiv:2502.09747*

Lim, S., & Schmälzle, R. (2024). The effect of source disclosure on evaluation of AI-generated messages. *Computers in Human Behavior: Artificial Humans, 2*(1), 100058.

Liu, G. (2022, June 21). The world's smartest artificial intelligence just made its first magazine cover. *Cosmopolitan*. https://www.cosmopolitan.com/lifestyle/a40314356/dall-e-2-artificial-intelligence-cover/

Mazzone, M., & Elgammal, A. (2019, February). Art, creativity, and the potential of artificial intelligence. *Arts 8*(1), 26.

Mello, J. (2024, May 1). Copyleaks study finds explosive growth in AI content on the web. *Tech News World*. https://www.technewsworld.com/story/copyleaks-study-finds-explosive-growth-of-ai-content-on-the-web-179161.html

Michalowski, M. (2024, November 29). How much data is generated every day in 2024? *Spacelift*. https://spacelift.io/blog/how-much-data-is-generated-every-day

Noy, S., & Zhang, W. (2023). Experimental evidence on the productivity effects of generative artificial intelligence. *Science, 381*, 187–192.

Park, J. Y. (2023). Could ChatGPT help you to write your next scientific paper?: Concerns on research ethics related to usage of artificial intelligence tools. *Journal of the Korean Association of Oral and Maxillofacial Surgeons, 49*(3), 105–106.

Pereira, D. (2024, March 6). By 2026, online content generated by non-humans will vastly outnumber human generated content. *Da Loop.* https://oodaloop.com/analysis/archive/if-90-of-online-content-will-be-ai-generated-by-2026-we-forecast-a-deeply-human-anti-content-movement-in-response/

Pinzolits, R. (2024). AI in academia: An overview of selected tools and their areas of application. *MAP Education and Humanities, 4*, 37–50.

Preiksaitis, C., & Rose, C. (2023). Opportunities, challenges, and future directions of generative artificial intelligence in medical education: Scoping review. *JMIR Medical Education, 9*, e48785.

Publishers Weekly. (2024, November 18). HarperCollins inks AI licensing deal for nonfiction books. https://www.publishersweekly.com/pw/newsbrief/index.html?record=5076#:~:text=HarperCollins%20has%20confirmed%20that%20they,company%20statement%20shared%20with%20PW

Rigney, D. (2010). *The Matthew effect: How advantage begets further advantage.* Columbia University Press.

Robledo, S., Grisales Aguirre, A. M., Hughes, M., & Eggers, F. (2023). "Hasta la vista, baby"–will machine learning terminate human literature reviews in entrepreneurship? *Journal of Small Business Management, 61*(3), 1314–1343.

Rosenberg, B. (2023). *"Whatever it is, I'm against it": Resistance to change in higher education.* Harvard Education Press.

Salah, M., Abdelfattah, F., & Halbusi, H. A. (2023). Debate: Peer reviews at the crossroads — "To AI or not to AI?" *Public Money & Management, 43*(8), 781–782.

Scherer, M. (2023, October 17). New WGA labor agreement gives Hollywood writers important protections in the era of AI. Center for Democracy & Technology. https://cdt.org/insights/new-wga-labor-agreement-gives-hollywood-writers-important-protections-in-the-era-of-ai/#:~:text=,writer%E2%80%99s%20credit%20or%20separated%20rights

Scopes, J. T., & Bryan, W. J. (1925). *The scopes trial.* Legal Classics Library.

Shapiro, A. R. (2016). The scopes trial. In J. Butler (Ed.), *Oxford research encyclopedia of American history.*

Sheikh, S. (2006). Notes on institutional critique. *Prelom, 8*, 217–219.

Siler, K. (2017). Future challenges and opportunities in academic publishing. *Cahiers canadiens de sociologie* [Canadian Journal of Sociology], *42*(1), 83–114.

Singha, R., Singha, S., & Jasmine, E. (2024). The intersection of academics and career readiness. In C. S. Conway & A. Jiahao Liu (Eds.), *Preparing students from the academic world to career paths: A comprehensive guide* (pp. 246–266). IGI Global.

Sturm, B. L., & Ben-Tal, O. (2021). Folk the algorithms: (Mis)applying artificial intelligence to folk music. In E. R. Miranda (Ed.), *Handbook of artificial intelligence for music: Foundations, advanced approaches, and developments for creativity* (pp. 423–454) Springer.

Taylor & Francis. (2023, February 17). Taylor & Francis clarifies the responsible use of AI tools in academic content creation. https://newsroom.taylorandfrancisgroup.com/taylor-francis-clarifies-the-responsible-use-of-ai-tools-in-academic-content-creation/#:~:text=Therefore%2C%20AI%20tools%20must%20not,be%20acknowledged%20and%20documented%20appropriately

Teixeira da Silva, J. A., Dobránszki, J., Bhar, R. H., & Mehlman, C. T. (2019). Editors should declare conflicts of interest. *Journal of Bioethical Inquiry, 16*, 279–298.

Tennant, J. (2020). Time to stop the exploitation of free academic labour. *European Science Editing, 46*, e51839.

Tennant, J. P., Waldner, F., Jacques, D. C., Masuzzo, P., Collister, L. B., & Hartgerink, C. H. (2016). The academic, economic and societal impacts of open access: An evidence-based review. *F1000Research, 5*, 632.

The Author's Guild. (2024, October 23). Author's Guild encouraged by Penguin Random House's new AI restrictions. https://authorsguild.org/news/ag-encouraged-by-penguin-random-house-ai-restrictions/#:~:text=The%20Authors%20Guild%20applauds%C2%A0Penguin%20Random,outs%20to%20use%20the%20books

TipRanks. (2025, February 17). *New York Times* (NYT) introduces new AI tools to help its newsroom. *Business Insider.* https://markets.businessinsider.com/news/stocks/new-york-times-nyt-introduces-new-ai-tools-to-help-its-newsroom-1034370803#:~:text=Media%20company%20New%20York%20Times,support%20and%20improve%20their%20work

Trumble, R., & van Riemsdijk, M. (2016). Commodification of art versus creativity: The Antagonist Art Movement in the expanding arts scene of New York City. *City, Culture and Society, 7*(3), 155–160.

Tuomi, I. (2024). Beyond mastery: Toward a broader understanding of AI in education. *International Journal of Artificial Intelligence in Education, 34*(1), 20–30.

Uddin, L., Harvey, D. M., Dafa, J. H., Trouvais, B. A., Harwell, E. N., Elga, B. D., & Herold, J. (2024). *Antitrust class action to challenge collusion among the world's six largest for-profit publishers of peer-reviewed scholarly journals.* https://digitalcommons.unl.edu/scholcom/330/

U.S. Copyright Office. (2025). *Copyright and artificial intelligence, Part 2: Copyrightability.* https://www.copyright.gov/ai

Varese, E., & Battistella, C. (2023, June 12). Can generative AI rely on the text and data mining (TDM) exception for its training? *DLA Piper.* https://www.dlapiper.com/es-pr/insights/publications/law-in-tech/l-ai-generativa-puo-fare-affidamento-sulla-eccezione-text-and-data-mining-per-il-suo-addestramento

Wagner, G., Lukyanenko, R., & Paré, G. (2022). Artificial intelligence and the conduct of literature reviews. *Journal of Information Technology, 37*(2), 209–226.

Walsh, D. (2023, October 3). Study gauges how people perceive AI-created content. MIT Management Sloan School. https://mitsloan.mit.edu/ideas-made-to-matter/study-gauges-how-people-perceive-ai-created-content

Willinsky, J. (2018). Scholarly associations and the economic viability of open access publishing. *Health Sciences: An OJS 3.4 Theme Demo, 1*(2) 180–183.

Zeichner, K. M. (1995). Beyond the divide of teacher research and academic research. *Teachers and Teaching, 1*(2), 153–172.

Žydžiūnaitė, V. (2016). Research area, work experience, and parents' completed higher education within scientists' intellectual leadership in higher education: Which roles matter? *European Scientific Journal, 12*(25), 9.

Embracing the Opaque

This concluding chapter confronts the ethical impasse surrounding transparency and disclosure in creative processes, advocating forcefully for creators' rights to opacity. Arguing against the widespread ethical presumption that creators have an inherent obligation to disclose their methods, it proposes instead a radical reframing: that creative methods—whether involving generative artificial intelligence (AI) or traditional techniques—should remain exclusively within the purview of the creators themselves. Not only is mandatory disclosure ethically questionable, but it is also practically unfeasible in environments saturated with artificial intelligence (AI) driven tools. Fine grained documentation required to "prove" purely human authorship quickly becomes impossible, rendering demands for transparency little more than performative rituals. Publishers retain authority over distribution methods; creators, in turn, must hold sovereign authority over their creative processes, judged ultimately on the quality and merit of their final products alone.

1 Toward a New Ethics of Creative Freedom

Throughout the preceding chapters, the exploration of creative autonomy versus institutional control has highlighted a fundamental tension: Creators increasingly find themselves beholden to demands for transparency and documentation that seek to regulate their methods. This pervasive expectation for disclosure has grown in prominence alongside advancements in generative AI tools, ostensibly as a safeguard against plagiarism, fraud, or a perceived loss of authenticity. Yet while such expectations may initially seem reasonable—even ethical—they ultimately undermine the essential autonomy upon which true creativity relies. An insistence on detailed disclosure paradoxically becomes not an ethical imperative but rather an intrusion, turning the private workshop of the artist or researcher into a space policed by institutional anxieties.

Consider the practical reality of creative work within digital ecosystems saturated with AI-driven tools. As argued throughout this volume, it is not merely inconvenient but wholly impractical to mandate the type of granular transparency demanded by disclosure proponents. How would a writer reliably document each interaction with generative software, down to subtle changes in syntax or the iterative conversations that refine an idea? As AI tools become seamlessly embedded into our workflows—much like spell-checkers, thesauruses, or search engines—it becomes increasingly nonsensical to separate human from machine contributions clearly. A demand for comprehensive disclosure of creative methods thus reveals itself as an unattainable standard, less a sincere pursuit of transparency than a reflexive reaction born of discomfort with technological evolution.

Further complicating the issue is the often-overlooked fact that creative work has never truly operated with absolute transparency; nor should it. Writers, artists, and scholars routinely engage in myriad opaque processes—drafting, editing, revising, discarding—that remain hidden behind final works. Historically, creators have enjoyed the freedom to select their tools without obligation to explain or justify those choices publicly. To impose detailed mandates now, simply because those tools involve AI, amounts to a selective ethics born out of suspicion and discomfort. Transparency, pursued without restraint, risks imposing a chilling effect on innovation, constraining creators into methods they believe will pass external scrutiny rather than those that best serve their creative visions.

This perspective is not an argument against accountability. Rather, it is an affirmation of trust in creators to manage their methodologies ethically without constant institutional oversight. Just as an instructor grades a student's final paper rather than their hours logged in a library or the precise notes they took, publishers and institutions should judge the quality and originality of work based solely on the outcome. The final creative product, not the intricacies of its creation, should be the definitive measure of its value. Such an approach restores respect and trust to the creative professional, reinforcing their autonomy and reinforcing creativity itself as an inherently independent human activity.

Indeed, creative autonomy does not merely support individual expression; it fundamentally underpins innovation. The history of literature, science, and art reveals that true breakthroughs often occur precisely because innovators diverge from established processes and methodologies. When publishers and institutions impose rigid methodological constraints—such as prohibitions on AI or demands for disclosure—they inadvertently stifle the conditions necessary for innovation. Creatives must remain free to experiment and even conceal their methods if they so choose, safe from interference and judgment. The power of creativity lies precisely in its unpredictability and in the artist's right to protect their methods, no matter how unconventional.

Critics of this stance argue that undisclosed use of generative tools constitutes an ethical breach—akin to deceit or plagiarism—based on a perceived purity of human-only creativity. Yet such views overlook the fundamental reality of how creativity has always operated. Human ingenuity has historically thrived through interactions with new technologies, from Gutenberg's printing press to photography, synthesizers, digital editing software, and now AI-driven text and image generation. None of these historical advancements required mandatory disclosure of their use; instead, creators freely integrated these tools into their workflows. The current insistence on disclosing AI involvement marks a striking departure from historical norms—one driven more by emotional discomfort with technological change than by objective ethical necessity.

In examining these emotional responses more closely, we observe that resistance often arises from deeper fears about authenticity, originality, and human value. An English professor's inflammatory analogy equating AI training methods with the enslavement of authors exemplifies this profoundly emotional reaction. Such charged rhetoric demonstrates less an objective ethical argument and more an intense anxiety about human displacement by machines. Rather than attempting to resolve these anxieties through mandates of transparency, it is more productive to acknowledge and engage with them openly. Understanding the emotional core of resistance helps clarify that the true challenge is not ethical disclosure but coming to terms with what it means to be human in collaboration with machines.

The creative community must redefine the ethics of authorship in this age of generative tools. A new ethical framework would reject absolute transparency as an ideal, embracing instead the concept of "ethical opacity." Ethical opacity recognizes the creator's right to choose their tools, protect their methods, and ultimately be judged only by the integrity and merit of the final product. Within this framework, ethics shift from obsessively policing methodology to evaluating outputs fairly and rigorously. Institutions, publishers, and creators alike would do well to internalize this principle, recalibrating their policies and attitudes accordingly.

There remains, of course, a distinction between policy and professional norms. Policies may officially allow generative tools, but professional communities—editors, reviewers, and academics—may continue to adhere to traditional practices by collective choice. Such divergence is perfectly legitimate; the argument presented throughout this volume explicitly supports the freedom of communities to define their own norms. However, it becomes problematic when groups attempt to universalize their norms, pressuring others who do not share their worldview. Just as a religious convert might zealously impose their beliefs onto unwilling recipients, so too does the creative community risk infringement upon others' freedom when dictating universal methodologies. Thus, the call for a new ethics of creative freedom is a rallying cry for independence and trust, an assertion that disclosure should never be mandated to satisfy institutional anxieties. Creators have no obligation—ethical or otherwise—to justify their processes to external observers. Publishers may rightly determine how content is disseminated, but creators alone must dictate how they create. To truly nurture creativity in an AI-integrated world, we must abandon impractical ideals of absolute transparency and embrace instead an ethical framework that respects autonomy, innovation, and, above all, the integrity of the creative individual.

2 The Future of Authorship in the AI Era

As this exploration of authorship, creativity, and institutional control draws to a close, one may feel that the ground beneath traditional assumptions has shifted irreversibly. This era of generative AI, far from just another passing technological fad, represents a transformation in the landscape of human creativity and intellectual labor. As unsettling as it may be, the entry of machines into our creative sanctuaries compels us to confront deeper anxieties and assumptions—about originality, human worth, and the very essence of authorship. The emotional reactions we see in debates over generative tools reveal how deeply these technologies strike at our core identities and values. The vehement opposition from some quarters, driven more by fear and discomfort than by objective ethical logic, underscores this point clearly.

Consider the deeply troubling analogy proposed by an English professor equating AI training methods with enslavement. According to this view, the forced training of AI on human-authored texts corrupts all subsequent outputs—implying that creativity and originality have been irreparably compromised, tainted by a kind of technological coercion. This comparison, as profoundly emotional and inflammatory as it is, reveals how high the stakes have become for those who feel their professional and personal identities are under threat. Such analogies invoke powerful emotions precisely because they touch upon deep-seated fears regarding autonomy, value, and the boundaries of human experience. Rational arguments about the inevitability of technological integration or historical precedent offer little consolation to those who perceive their roles, identities, or dignity as at risk.

Yet, as compelling as emotional reactions are, they rarely represent reliable guides for ethical or practical policy. Research in cognitive psychology and persuasion has consistently shown that humans are rarely swayed by facts alone—especially when deeply held beliefs or identities feel threatened (Haidt, 2012; Mercier & Sperber, 2017). This book, too, acknowledges that facts alone may not have changed minds; for many, the emotional responses against AI's integration into creative processes run deeper than logical reasoning or historical analogies can soothe. However, identifying the emotions underlying these intense reactions—fear, anger, suspicion—may help facilitate more constructive dialogue and thoughtful introspection.

Ultimately, the question facing creators, institutions, and society at large is not merely how to regulate these powerful new technologies but how we choose to coexist with them. The integration of generative AI is already well underway, permeating every creative domain from journalism and academia to visual art and film production. Trying to hold back this tide through restrictive disclosure mandates or outright bans is akin to halting industrialization or the printing press. Instead, our challenge is learning to thoughtfully embrace these tools, understanding their potentials, limitations, and—critically—our own evolving relationship to them.

In fact, the AI era provides a unique opportunity for introspection and self-awareness within creative communities. As we grapple with what precisely unsettles us about sharing creative labor with machines, we discover truths about what we value most deeply: our humanity, our identity, and our cultural ideals of originality and authenticity. The fear that machines diminish human contributions reveals an insecurity about our collective worth and legacy. Yet, throughout history, human creativity has consistently risen above technological disruptions, adapting to new tools and paradigms, and ultimately enriching culture rather than diminishing it.

Our vision of authorship in the future must therefore evolve beyond binary oppositions of human-versus-machine. Instead, we must imagine hybridized modes of creation—centaur and cyborg collaborations—that extend rather than undermine human capabilities. What we risk losing through reactionary responses is the potential for profound innovation, enriched human expression, and expanded creative horizons. Embracing generative tools does not entail relinquishing human agency or authenticity; rather, it represents the next logical step in humanity's ongoing dialogue with technology.

This book has passionately defended the radical idea that creators should possess full autonomy over their methods, tools, and processes. Such freedom means preserving the right to opacity — resisting unnecessary intrusion by institutional or ethical policing. Just as we trust the final product of a student's scholarship or an artist's performance, so too must we trust creative professionals in their chosen methods of production. Accountability is found not in exhaustive documentation or disclosure but in the inherent quality, originality, and integrity of the final creative output.

Indeed, the future of authorship in the AI era demands a reconsideration of traditional ethical assumptions — moving away from mandatory transparency toward ethical opacity and moving away from policing methods toward evaluating products. This approach may provoke discomfort or resistance from those accustomed to older norms. Yet it represents the only sustainable path forward in an era where human and machine labor become indistinguishable parts of the creative process. Trust in creators, supported by policies affirming autonomy and respect for diversity of method, can sustain vibrant innovation even amid rapid technological change.

Looking forward, we must challenge ourselves not merely to coexist with generative AI but also to thrive alongside it. The invitation now is toward introspection and courageous visioning. What deeply held assumptions about creativity and authorship are we clinging to out of fear or inertia? How might embracing uncertainty and redefining our boundaries enrich rather than impoverish our creative culture? And perhaps most provocatively, what new artistic, intellectual, and cultural horizons might unfold if we loosen our grip on outdated ideas of authenticity and purity, allowing ourselves instead to be co-creators with machines?

This horizon — unknown, uncertain, yet full of transformative potential — is the challenge before us. It demands introspection, flexibility, and perhaps a bit of courage. The future of authorship in the AI era will not be determined by machines alone, or by reactionary institutional policies, but by our willingness to engage openly, thoughtfully, and fearlessly with a world in which creativity itself is evolving. The invitation now extends to each reader to reflect deeply on these questions, to challenge one's own emotional boundaries, and to envision a future where humans and machines collaborate not as competitors but as co-creators, boldly charting unexplored realms of imaginative possibility.

References

Haidt, J. (2013). Moral psychology for the twenty-first century. Journal of Moral Education, 42(3), 281–297.

Mercier, H., & Sperber, D. (2017). The enigma of reason. Harvard University Press.

INDEX

CONTENTS